Carlos Bush

Inside Jupiter

Contents

1.

2.

    1.

    2.

    3.

      4.

      5.

      6.

      7.

8.
9.
10.
11.
12.
13.
14.
15.
16.
17.
18.
19.
20.
21.
22.
23.
24.
25.
26.
27.
3.
1.
2.
3.

4.

5.

6.

7.

8.

9.

10.

Foreword

July 5, 2016

2

Astronomy & Space Space Exploration

Welcome to Jupiter: NASA spacecraft Juno reaches giant planet

by Alicia Chang

This artist's rendering provided by NASA and JPL-Caltech shows the Juno spacecraft above the planet Jupiter. Five years after its launch from Earth, Juno is scheduled to go into orbit around the gas giant on Monday, July 4, 2016. (NASA/JPL-Caltech via AP)

Braving intense radiation, a NASA spacecraft reached Jupiter on Monday after a five-year voyage to begin exploring the king of the planets.

Ground controllers at the NASA Jet Propulsion Laboratory and Lockheed Martin erupted in applause when the solar-powered Juno spacecraft beamed home news that it was circling Jupiter's poles.

The arrival at Jupiter was dramatic. As Juno approached its target, it fired its rocket engine to slow itself down and gently slipped into orbit. Because of the communication time lag between Jupiter and Earth, Juno was on autopilot when it executed the tricky move.

"Juno, welcome to Jupiter," said mission control commentator Jennifer Delavan of Lockheed Martin, which built Juno.

Mission managers said early reports indicated Juno was healthy and performed flawlessly.

"Juno sang to us and it was a song of perfection," JPL project manager Rick Nybakken said during a post-mission briefing.

The spacecraft's camera and other instruments were switched off for arrival, so there weren't any pictures at the moment it reached its destination. Afterward, NASA released a time-lapse video taken last week during the approach, showing Jupiter glowing yellow in the distance and its four inner moons dancing around it.

The view yielded a surprise: Jupiter's second-largest moon, Callisto, appeared dimmer than initially thought. Scientists have promised close-up views of the planet when Juno skims the cloud tops during the 20-month, $1.1 billion mission.

The fifth rock from the sun and the heftiest planet in the solar system, Jupiter is what's known as a gas giant—a ball of hydrogen and helium—unlike rocky Earth and Mars.

With its billowy clouds and colorful stripes, Jupiter is an extreme world that likely formed first, shortly after the sun. Unlocking its history may hold clues to understanding how Earth and the rest of the solar system developed.

Named after Jupiter's cloud-piercing wife in Roman mythology, Juno is only the second mission designed to spend time at Jupiter.

Galileo, launched in 1989, circled Jupiter for nearly a decade, beaming back splendid views of the planet and its numerous moons. It uncovered signs of an ocean beneath the icy surface of the moon Europa, considered a top target in the search for life outside Earth.

Juno's mission: To peer through Jupiter's cloud-socked atmosphere and map the interior from a unique vantage point above the poles. Among the lingering questions: How much water exists? Is there a solid core? Why are Jupiter's southern and northern lights the brightest in the solar system?

"What Juno's about is looking beneath that surface," Juno chief scientist Scott Bolton said before the arrival. "We've got to go down and look at what's inside, see how it's built, how deep these features go, learn about its real secrets."

There's also the mystery of its Great Red Spot. Recent observations by the Hubble Space Telescope revealed the centuries-old monster storm in Jupiter's atmosphere is shrinking.

The trek to Jupiter, spanning nearly five years and 1.8 billion miles (2.8 billion kilometers), took Juno on a tour of the inner solar system followed by a swing past Earth that catapulted it beyond the asteroid belt between Mars and Jupiter.

Along the way, Juno became the first spacecraft to cruise that far out powered by the sun, beating Europe's comet-chasing Rosetta spacecraft. A trio of massive solar wings sticks out from Juno like blades from a windmill, generating 500 watts of power to run its nine instruments.

In the coming days, Juno will turn its instruments back on, but the real work won't begin until late August when the spacecraft swings in closer. Plans called for Juno to swoop within 3,000 miles (5,000 kilometers) of Jupiter's clouds—closer than previous missions—to map the planet's gravity and magnetic fields in order to learn about the interior makeup.

Juno is an armored spacecraft—its computer and electronics are locked in a titanium vault to shield them from harmful radiation. Even so, Juno is expected to get blasted with radiation equal to more than 100 million dental X-rays during the mission.

Like Galileo before it, Juno meets its demise in 2018 when it deliberately dives into Jupiter's atmosphere and disintegrates—a necessary sacrifice to prevent any chance of accidentally crashing into the planet's potentially habitable moons.

1

Inside Jupiter

Jupiter is the fifth planet from the Sun and the largest in the Solar System. It is a gas giant with a mass more than two and a half times that of all the other planets in the Solar System combined, but slightly less than one-thousandth the mass of the Sun. Jupiter is the third brightest natural object in the Earth's night sky after the Moon and Venus, and it has been observed since prehistoric times. It was named after the Roman god Jupiter, the king of the gods.

Jupiter

Full disk view of Jupiter, taken by the Hubble Space Telescope in 2020[a]

Designations

Pronunciation

/ˈdʒuːpɪtər/ (

🔊

listen)[1]

Named after

Jupiter

Adjectives

Jovian /ˈdʒoʊviən/

Orbital characteristics[7]

Epoch J2000

Aphelion

816.363 Gm(5.4570 AU)

Perihelion

740.595 Gm (4.9506 AU)

Semi-major axis

778.479 Gm (5.2038 AU)

Eccentricity

0.0489

Orbital period (sidereal)

- 11.862 yr
- 4,332.59 d
- 10,476.8 Jovian solar days[2]

Orbital period (synodic)

398.88 d

Average orbital speed

13.07 km/s (8.12 mi/s)

Mean anomaly

20.020°[3]

Inclination

- 1.303° to ecliptic[3]
- 6.09° to Sun's equator[3]
- 0.32° to invariable plane[4]

Longitude of ascending node

100.464°

Time of perihelion

21 January 2023[5]

Argument of perihelion

273.867°[3]

Known satellites

80 (as of 2021)[6]

Physical characteristics[7][13][14]

Mean radius

69,911 km (43,441 mi)[b]

10.973 of Earth's

Equatorial radius

71,492 km (44,423 mi)[b]

11.209 of Earth's

Polar radius

66,854 km (41,541 mi)[b]

10.517 of Earth's

Flattening

0.06487

6.1469×10^{10} km^2(2.3733×10^{10} sq mi)

120.4 of Earth's

Volume

1.4313×10^{15} km^3(3.434×10^{14} cu mi)[b]

1,321 of Earth's

Mass

1.8982×10^{27} kg (4.1848×10^{27} lb)

- 317.8 of Earth's
- 1/1047 of Sun's[8]

Mean density

1,326 kg/m3(2,235 lb/cu yd)[c]

Surface gravity

24.79 m/s2 (81.3 ft/s2)[b]

2.528 g

Moment of inertia factor

0.2756±0.0006[9]

Escape velocity

59.5 km/s (37.0 mi/s)[b]

Synodicrotation period

9.9258 h (9 h 55 m 33 s)[2]

Sidereal rotation period

9.9250 hours (9 h 55 m 30 s)

Equatorial rotation velocity

12.6 km/s (7.8 mi/s; 45,000 km/h)

Axial tilt

3.13° (to orbit)

North pole right ascension

268.057°; 17^h 52^m 14^s

North pole declination

64.495°

Albedo

0.503 (Bond)[10]

0.538 (geometric)[11]

Surface temp.minmeanmax

1 bar

165 K

0.1 bar

78 K

128 K

Apparent magnitude

−2.94[12] to −1.66[12]

Angular diameter

29.8" to 50.1"

Atmosphere[7]

Surface pressure

200–600 kPa (opaque cloud deck)[15]

Scale height

27 km (17 mi)

Composition by volume

- 89%±2.0% hydrogen(H_2)
- 10%±2.0% helium(He)
- 0.3%±0.1% methane(CH_4)
- 0.026%±0.004%ammonia (NH_3)
- 0.0028%±0.001%hydrogen deuteride(HD)
- 0.0006%±0.0002%ethane (C_2H_6)
- 0.0004%±0.0004%water (H_2O)

Jupiter is primarily composed of hydrogen, but helium constitutes one-quarter of its mass and one-tenth of its volume. It probably has a rocky core of heavier elements,[16] but, like the other giant planets in the Solar System, it lacks a well-defined solid surface. The ongoing contraction of Jupiter's interior generates more heat than it receives from the Sun. Because of its rapid rotation, the planet's shape is an oblate spheroid: it has a slight but noticeable bulge around the equator. The outer atmosphere is divided into a series of latitudinal bands, with turbulence and storms along their interacting boundaries. A prominent result of this is the Great Red Spot, a giant storm which has been observed since at least 1831.

Jupiter is surrounded by a faint planetary ringsystem and a powerful magnetosphere. Jupiter's magnetic tail is nearly 800 million km(5.3 AU; 500 million mi) long, covering nearly the entire distance to Saturn's orbit. Jupiter has 80 known moons and possibly many more,[6] including the four large moons discovered by Galileo Galilei in 1610: Io, Europa, Ganymede, and Callisto. Io and Europa are about the size of Earth's Moon; Callisto is almost the size of the planet Mercury, and Ganymede is larger.

Pioneer 10 was the first spacecraft to visit Jupiter, making its closest approach to the planet in December 1973.[17] Jupiter has since been explored by multiple robotic spacecraft, beginning with the Pioneer and Voyager flybymissions from 1973 to 1979, and later with the *Galileo*orbiter in 1995.[18] In 2007, the New Horizons visited Jupiter using its gravity to increase its speed, bending its trajectory en route to Pluto. The latest probe to visit the planet, Juno, entered orbit around Jupiter in July 2016.[19][20] Future targets for exploration in the Jupiter system include the probable ice-covered liquid ocean of Europa.[21]

Name and symbol

In both the ancient Greek and Roman civilizations, Jupiter was named after the chief god of the divine pantheon: Zeus for the Greeks and Jupiter for the Romans. The International Astronomical Union (IAU) formally adopted the name Jupiter for the planet in 1976. The IAU names newly discovered satellites of Jupiter for the mythological lovers, favorites, and descendants of the god.[22]The planetary symbol for Jupiter,

♃

, descends from a Greek zeta with a horizontal stroke, ⟨Ƶ⟩, as an abbreviation for *Zeus*.[23][24]

Jove, the archaic name of Jupiter, came into use as a poetic name for the planet around the 14th century.[25] The Romans named the fifth day of the week *diēsIovis* ("Jove's Day") after the planet Jupiter.[26] In Germanic mythology, Jupiter is equated to Thor, whence the English name *Thursday* for the Roman *dies Jovis*.[27]

The original Greek deity *Zeus* supplies the root *zeno-*, which is used to form some Jupiter-related words, such as zenographic.[d] *Jovian*is the adjectival form of Jupiter. The older adjectival form *jovial*, employed by astrologers in the Middle Ages, has come to mean "happy" or "merry", moods ascribed to Jupiter's astrological influence.[28]

Formation and migration

Main article: Grand tack hypothesis

See also: Formation and evolution of the Solar System

Jupiter is believed to be the oldest planet in the Solar System.[29] Current models of Solar System formation suggest that Jupiter formed at or beyond the snow line: a distance from the early Sun where the temperature is sufficiently cold for volatiles such as water to condense into solids.[30] The planet began as a large solid core, then accumulated its gaseous atmosphere. As a consequence, the core must have formed before the solar nebula was fully dispersed after 10 million years. Over about a million years, Jupiter's atmosphere gradually expanded until it had 20 times the mass of the Earth. The orbiting mass created a gap in the solar nebula, and thereafter the planet slowly increased to 50 Earth masses over 3–4 million years.[29]

According to the "grand tack hypothesis", Jupiter began to form at a distance of roughly 3.5 AU (520 million km; 330 million mi) from the Sun. As the young planet accreted mass, interaction with the gas disk orbiting the Sun and orbital resonances with Saturn caused it to migrate inward.[30][31] This upset the orbits of several super-Earths orbiting closer to the Sun, causing them to collide destructively. Saturn would later have begun to migrate inwards too, much faster than Jupiter, until the two planets became captured in a 3:2 mean motion resonance at approximately 1.5 AU (220 million km; 140 million mi) from the Sun. This changed the direction of migration, causing them to migrate away from the Sun and out of the inner system to their current locations.[32] All of this happened over a period of 3–6 million years, with the final migration of Jupiter occurring over several hundred thousand years.[31][33] Jupiter's departure from the inner solar system eventually allowed the inner planets—including Earth—to form from the rubble.[34]

There are several problems with the grand tack hypothesis. The resulting formation timescales of terrestrial planets appear to be inconsistent with the measured elemental composition.[35]It is likely that Jupiter would have settled into an orbit much closer to the Sun if it had migrated through the solar nebula.[36] Some competing models of Solar System formation predict the formation of Jupiter with orbital properties that are close to those of the present day planet.[37] Other models predict Jupiter forming at distances much farther out, such as 18 AU (2.7 billion km; 1.7 billion mi).[38][39]

Based on Jupiter's composition, researchers have made the case for an initial formation outside the molecular nitrogen (N_2) snowline, which is estimated at 20–30 AU (3.0–4.5 billion km; 1.9–2.8 billion mi) from the Sun,[40][41] and possibly even outside the argon snowline, which may be as far as 40 AU (6.0 billion km; 3.7 billion mi). Having formed at one of these extreme distances, Jupiter would then have migrated inwards to its current location. This inward migration would have occurred over a roughly 700,000-year time period,[38][39] during an epoch approximately 2–3 million years after the planet began to form. In this model, Saturn, Uranus and Neptune would have formed even further out than Jupiter, and Saturn would also have migrated inwards.

Physical characteristics

Jupiter is a gas giant, being primarily composed of gas and liquid rather than solid matter. It is the largest planet in the Solar System, with a diameter of 142,984 km (88,846 mi) at its equator.[42] The average density of Jupiter, 1.326 g/cm^3, is about the same as simple syrup (syrup USP),[43] and is lower than those of the four terrestrial planets.[44][45]

Composition

Jupiter's upper atmosphere is about 90% hydrogen and 10% helium by volume. Since helium atoms are more massive than hydrogen molecules, Jupiter's atmosphere is approximately 24% helium by mass.[46] The atmosphere contains trace amounts of methane, water vapour, ammonia, and silicon-based compounds. There are also fractional amounts of carbon, ethane, hydrogen sulfide, neon, oxygen, phosphine, and sulfur. The outermost layer of the atmosphere contains crystals of frozen ammonia. Through infraredand ultraviolet measurements, trace amounts of benzene and other hydrocarbons have also been found.[47] The interior of Jupiter contains denser materials—by mass it is roughly 71% hydrogen, 24% helium, and 5% other elements.[48][49]

The atmospheric proportions of hydrogen and helium are close to the theoretical composition of the primordial solar nebula. Neon in the upper atmosphere only consists of 20 parts per million by mass, which is about a tenth as abundant as in the Sun.[50] Helium is also reduced to about 80% of the Sun's helium composition. This depletion is a result of precipitation of these elements as helium-rich droplets, a process that happens deep in the interior of the planet.[51][52]

Based on spectroscopy, Saturn is thought to be similar in composition to Jupiter, but the other giant planets Uranus and Neptune have relatively less hydrogen and helium and relatively more of the next most common elements, including oxygen, carbon, nitrogen, and sulfur.[53] These planets are known as ice giants, because the majority of their volatilecompounds are in solid form.

Size and mass

Main article: Jupiter mass

Jupiter with its moon Europa on the left. Earth's diameter is 11 times smaller than Jupiter, and 4 times larger than Europa.

Jupiter's mass is 2.5 times that of all the other planets in the Solar System combined—so massive that its barycentre with the Sun lies above the Sun's surface at 1.068 solar radiifrom the Sun's centre.[54] Jupiter is much larger than Earth and considerably less dense: it has 1,321 times the volume of the Earth, but only 318 times the mass.[7][55]:6 Jupiter's radius is about one tenth the radius of the Sun,[56] and its mass is one thousandth the mass of the Sun, as the densities of the two bodies are similar.[57] A "Jupiter mass" (M_J or M_{Jup}) is often used as a unit to describe masses of other objects, particularly extrasolarplanetsand brown dwarfs. For example, the extrasolar planet HD209458 b has a mass of 0.69 M_J, while Kappa Andromedaeb has a mass of 12.8 M_J.[58]

Theoretical models indicate that if Jupiter had over 40% more mass, the interior would be so compressed that its volume would *decrease*despite the increasing amount of matter. For smaller changes in its mass, the radius would not change appreciably.[59] As a result, Jupiter is thought to have about as large a diameter as a planet of its composition and evolutionary history can achieve.[60] The process of further shrinkage with increasing mass would continue until appreciable stellar ignition was achieved.[61] Although Jupiter would need to be about 75 times more massive to fuse hydrogen and become a star,[62] the smallest red dwarf may be only slightly larger in radius than Saturn.[63]

Jupiter radiates more heat than it receives through solar radiation, due to the Kelvin–Helmholtz mechanism within its contracting interior.[64]:30[65] This process causes Jupiter to shrink by about 1 mm (0.039 in)/yr.[66][67]When it formed, Jupiter was hotter and was about twice its current diameter.[68]

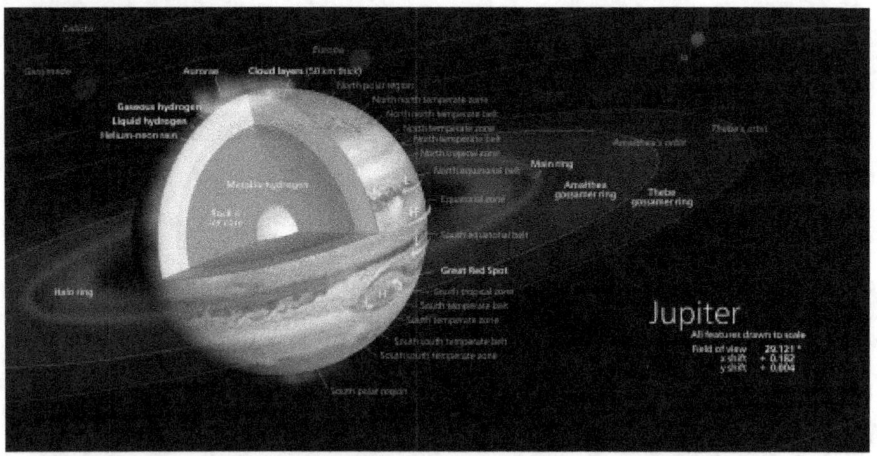

Diagram of Jupiter, its interior, surface features, rings, and inner moons.

Before the early 21st century, most scientists proposed one of two scenarios for the formation of Jupiter. If the planet accreted first as a solid body, it would consist of a dense core, a surrounding layer of liquid metallic hydrogen (with some helium) extending outward to about 80% of the radius of the planet,[69] and an outer atmosphere consisting primarily of molecular hydrogen.[67]Alternatively, if the planet collapsed directly from the gaseous protoplanetarydisk, it was expected to completely lack a core, consisting instead of denser and denser fluid (predominantly molecular and metallic hydrogen) all the way to the centre. Data from the *Juno*mission showed that Jupiter has a very diffuse core that mixes into its mantle.[19][70][71] This could have been caused by an impact from a planet of about ten Earth masses a few million years after Jupiter's formation, which would have disrupted an originally solid Jovian core.[72][73] It is estimated that the core takes up 30–50% of the planet's radius, and contains heavy elements with a combined mass 7–25 times the Earth.[74]

Outside the layer of metallic hydrogen lies a transparent interior atmosphere of hydrogen. At this depth, the pressure and temperature are above molecular hydrogen's critical pressure of 1.3 MPa and critical temperature of 33 K (−240.2 °C; −400.3 °F).[75] In this state, there are no distinct liquid and gas phases—hydrogen is said to be in a supercritical fluidstate. The hydrogen and helium gas extending downward from the cloud layer gradually transitions to a liquid in deeper layers, possibly resembling something akin to an ocean of liquid hydrogen and other supercritical fluids.[64]:22[76][77][78] Physically, the gas gradually becomes hotter and denser as depth increases.[79][80]

Rain-like droplets of helium and neon precipitate downward through the lower atmosphere, depleting the abundance of these elements in the upper atmosphere.[51][81]Calculations suggest that helium drops separate from metallic hydrogen at a radius of 60,000 km (37,000 mi) (11,000 km (6,800 mi) below the cloud tops) and merge again at 50,000 km (31,000 mi) (22,000 km (14,000 mi) beneath the clouds).[82] Rainfalls of diamond shave been suggested to occur, as well as on Saturn[83] and the ice giants Uranus and Neptune.[84]

The temperature and pressure inside Jupiter increase steadily inward because the heat of planetary formation can only escape by convection.[52] At a surface depth where the atmospheric pressure level is 1 bar (0.10 MPa), the temperature is around 165 K (−108 °C; −163 °F). The region of supercritical hydrogen changes gradually from a molecular fluid to a metallic fluid spans pressure ranges of

50–400 GPa with temperatures of 5,000–8,400 K (4,730–8,130 °C; 8,540–14,660 °F), respectively. The temperature of Jupiter's diluted core is estimated to be 20,000 K (19,700 °C; 35,500 °F) with a pressure of around 4,000 GPA.[85]

Atmosphere

Main article: Atmosphere of Jupiter

The atmosphere of Jupiter extends to a depth of 3,000 km (2,000 mi) below the cloud layers.[85]

Cloud layers

View of Jupiter's south pole

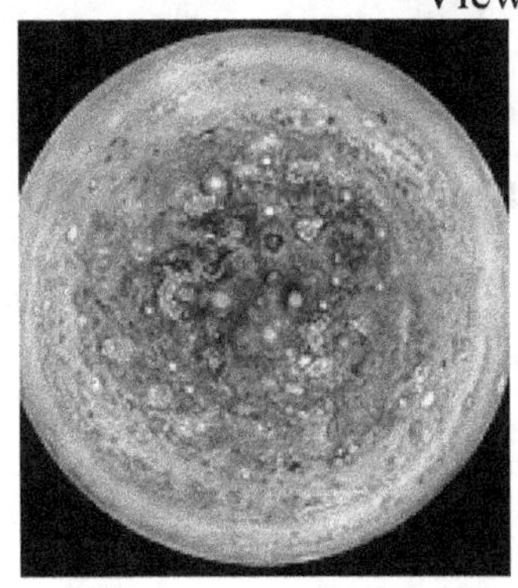

Enhanced color view of Jupiter's southern storms

Jupiter is perpetually covered with clouds of ammonia crystals, which may contain ammonium hydro sulfide as well.[86] The clouds are located in the tropopause layer of the atmosphere, forming bands at different latitudes, known as tropical regions. These are subdivided into lighter-hued *zones* and darker *belts*. The interactions of these conflicting circulation patterns cause storms and turbulence. Wind speeds of 100 metres per second (360 km/h; 220 mph) are common in zonal jet streams.[87] The zones have been observed to vary in width, colour and intensity from year to year, but they have remained stable enough for scientists to name them.[55]:6

The cloud layer is about 50 km (31 mi) deep, and consists of at least two decks of ammonia clouds: a thin clearer region on top with a thick lower deck. There may be a thin layer of waterclouds underlying the ammonia clouds, as suggested by flashes of lightning detected in the atmosphere of Jupiter.[88] These electrical discharges can be up to a thousand times as powerful as lightning on Earth.[89] The water clouds are assumed to generate thunderstorms in the same way as terrestrial thunderstorms, driven by the heat rising from the interior.[90] The Juno mission revealed the presence of "shallow lightning" which originates from ammonia-water clouds relatively high in the atmosphere.[91] These discharges carry "mushballs" of water-ammonia slushes covered in ice, which fall deep into the atmosphere.[92] Upper-atmospheric lightning has been observed in Jupiter's upper atmosphere, bright flashes of light that last around 1.4 milliseconds. These are known as "elves" or "sprites" and appear blue or pink due to the hydrogen.[93][94]

The orange and brown colours in the clouds of Jupiter are caused by upwelling compounds that change colour when they are exposed to ultraviolet light from the Sun. The exact makeup remains uncertain, but the substances are thought to be made up of phosphorus, sulfur or possibly hydrocarbons.[64]:39[95]These colourful compounds, known as chromophores, mix with the warmer clouds of the lower deck. The light-coloured zones are formed when rising convection cells form crystallising ammonia that hides the chromophores from view.[96]

Jupiter's low axial tilt means that the poles always receive less solar radiation than the planet's equatorial region. Convection within the interior of the planet transports energy to the poles, balancing out the temperatures at the cloud layer.[55]:54

Great Red Spot and other vortices

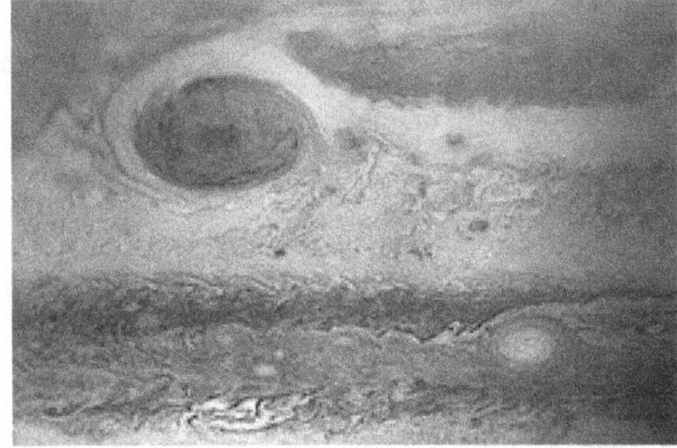

Close up of the Great Red Spot imaged by the *Juno*spacecraft in April 2018

The best known feature of Jupiter is the Great Red Spot,[97] a persistent anticyclonic storm located 22° south of the equator. It is known to have existed since at least 1831,[98] and possibly since 1665.[99][100] Images by the Hubble Space Telescope have shown as many as two "red spots" adjacent to the Great Red Spot.[101][102] The storm is visible through Earth-based telescopes with an aperture of 12 cm or larger.[103] The oval object rotates counterclockwise, with a period of about six days.[104] The maximum altitude of this storm is about 8 km (5 mi) above the surrounding cloudtops.[105] The Spot's composition and the source of its red colour remain uncertain, although photodissociated ammonia reacting with acetylene is a likely explanation.[106]

The Great Red Spot is larger than the Earth.[107] Mathematical models suggest that the storm is stable and will be a permanent feature of the planet.[108] However, it has significantly decreased in size since its discovery. Initial observations in the late 1800s showed it to be approximately 41,000 km (25,500 mi) across. By the time of the Voyager flybys in 1979, the storm had a length of 23,300 km (14,500 mi) and a width of approximately 13,000 km (8,000 mi).[109] *Hubble* observations in 1995 showed it had decreased in size to 20,950 km (13,020 mi), and observations in 2009 showed the size to be 17,910 km (11,130 mi). As of 2015, the storm was measured at approximately 16,500 by 10,940 km (10,250 by 6,800 mi),[109] and was decreasing in length by about 930 km (580 mi) per year.[107][110] In October 2021, a *Juno* flyby mission measured the depth of the Great Red Spot, putting it at around 300–500 kilometres (190–310 mi).[111]

Juno missions show that there are several polar cyclone groups at Jupiter's poles. The northern group contains nine cyclones, with a large one in the centre and eight others around it, while its southern counterpart also consists of a centre vortex but is surrounded by five large storms and a single smaller one.[112][113]These polar structures are caused by the turbulence in Jupiter's atmosphere and can be compared with the hexagon at Saturn's north pole.

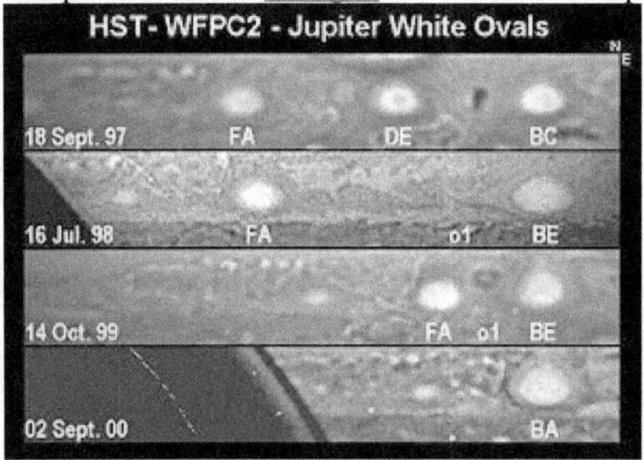

Formation of Oval BA from three white ovals

In 2000, an atmospheric feature formed in the southern hemisphere that is similar in appearance to the Great Red Spot, but smaller. This was created when smaller, white oval-shaped storms merged to form a single feature—these three smaller white ovals were formed in 1939–1940. The merged feature was named Oval BA. It has since increased in intensity and changed from white to red, giving it the nickname "Little Red Spot".[114][115]

In April 2017, a "Great Cold Spot" was discovered in Jupiter's thermosphere at its north pole. This feature is 24,000 km (15,000 mi) across, 12,000 km (7,500 mi) wide, and 200 °C (360 °F) cooler than surrounding material. While this spot changes form and intensity over the short term, it has maintained its general position in the atmosphere for more than 15 years. It may be a giant vortex similar to the Great Red Spot, and appears to be quasi-stable like the vortices in Earth's thermosphere. This feature may be formed by interactions between charged particles generated from Io and the strong magnetic field of Jupiter, resulting in a redistribution of heat flow.[116]

Magnetosphere

Main article: Magnetosphere of Jupiter

Aurorae on the north and south poles
(animation)

Aurorae on the north pole
(Hubble)

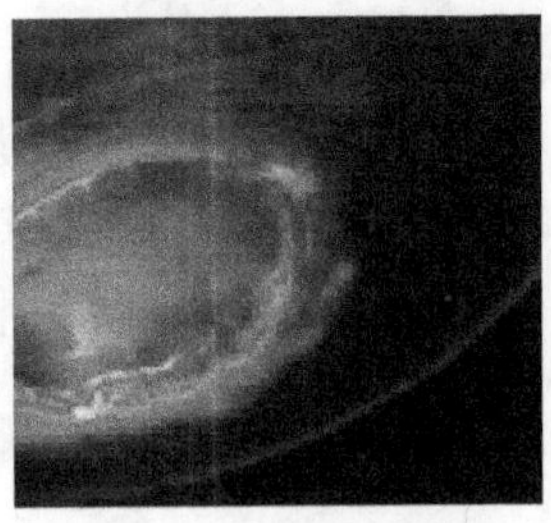

Infrared view of southern lights
(Jovian IR Mapper)

Jupiter's magnetic field is the strongest of any planet in the Solar System,[96] with a dipole moment of 4.170 gauss (0.4170 mT) that is tilted at an angle of 10.31° to the pole of rotation. The surface magnetic field strength varies from 2 gauss (0.20 mT) up to 20 gauss (2.0 mT).[117] This field is thought to be generated by eddy currents—swirling movements of conducting materials—within the liquid metallic hydrogen core. At about 75 Jupiter radii from the planet, the interaction of the magnetosphere with the solar windgenerates a bow shock. Surrounding Jupiter's magnetosphere is a magnetopause, located at the inner edge of a magnetosheath—a region between it and the bow shock. The solar wind interacts with these regions, elongating the magnetosphere on Jupiter's lee side and extending it outward until it nearly reaches the orbit of Saturn. The four largest moons of Jupiter all orbit within the magnetosphere, which protects them from the solar wind.[64]:69

The volcanoes on the moon Io emit large amounts of sulfur dioxide, forming a gas torusalong the moon's orbit. The gas is ionized in Jupiter's magnetosphere, producing sulfur and oxygen ions. They, together with hydrogen ions originating from the atmosphere of Jupiter, form a plasma sheet in Jupiter's equatorial plane. The plasma in the sheet co-rotates with the planet, causing deformation of the dipole magnetic field into that of a magnetodisk. Electrons within the plasma sheet generate a strong radio signature, with short, superimposed bursts in the range of 0.6–30 MHz that are detectable from Earth with consumer-grade shortwave radio receivers.[118][119] As Io moves through this torus, the interaction generates Alfvén waves that carry ionized matter into the polar regions of Jupiter. As a result, radio waves are generated through a cyclotron maser mechanism, and the energy is transmitted out along a cone-shaped surface. When Earth intersects this cone, the radio emissions from Jupiter can exceed the radio output of the Sun.[120]

Planetary rings

Main article: Rings of Jupiter

Jupiter has a faint planetary ring system composed of three main segments: an inner torus of particles known as the halo, a relatively bright main ring, and an outer gossamer ring.[121] These rings appear to be made of dust, while Saturn's rings are made of ice.[64]:65 The main ring is most likely made out of material ejected from the satellites Adrasteaand Metis, which is drawn into Jupiter because of the planet's strong gravitational influence. New material is added by additional impacts.[122] In a similar way, the moons Thebe and Amalthea are believed to produce the two distinct components of the dusty gossamer ring.[122] There is evidence of a fourth ring that may consist of collisional debris from Amalthea that is strung along the same moon's orbit.[123]

Orbit and rotation

Orbit of Jupiter and other outer Solar System planets

Jupiter is the only planet whose barycentrewith the Sun lies outside the volume of the Sun, though by only 7% of the Sun's radius.[124][125][126] The average distance between Jupiter and the Sun is 778 million km (5.2 AU) and it completes an orbit every 11.86 years. This is approximately two-fifths the orbital period of Saturn, forming a near orbital resonance.[127] The orbital plane of Jupiter is inclined 1.30° compared to Earth. Because the eccentricity of its orbit is 0.049, Jupiter is slightly over 75 million km nearer the Sun at perihelion than aphelion.[7]

The axial tilt of Jupiter is relatively small, only 3.13°, so its seasons are insignificant compared to those of Earth and Mars.[128]

Jupiter's rotation is the fastest of all the Solar System's planets, completing a rotation on its axis in slightly less than ten hours; this creates an equatorial bulge easily seen through an amateur telescope. Because Jupiter is not a solid body, its upper atmosphere undergoes differential rotation. The rotation of Jupiter's polar atmosphere is about 5 minutes longer than that of the equatorial atmosphere.[129]The planet is an oblate spheroid, meaning that the diameter across its equator is longer than the diameter measured between its poles.[80]On Jupiter, the equatorial diameter is 9,276 km (5,764 mi) longer than the polar diameter.[7]

Three systems are used as frames of reference for tracking the planetary rotation, particularly when graphing the motion of atmospheric features. System I applies to latitudes from 7° N to 7° S; its period is the planet's shortest, at 9h 50m 30.0s. System II applies at latitudes north and south of these; its period is 9h 55m 40.6s.[130] System III was defined by radio astronomers and corresponds to the rotation of the planet's magnetosphere; its period is Jupiter's official rotation.[131]

Observation

Jupiter and four Galilean moons seen through an amateur telescope

Jupiter is usually the fourth brightest object in the sky (after the Sun, the Moon, and Venus),[96] although at opposition Mars can appear brighter than Jupiter. Depending on Jupiter's position with respect to the Earth, it can vary in visual magnitude from as bright as −2.94 at opposition down to −1.66 during conjunctionwith the Sun.[12] The mean apparent magnitude is −2.20 with a standard deviation of 0.33.[12]The angular diameter of Jupiter likewise varies from 50.1 to 30.5 arc seconds.[7] Favourable oppositions occur when Jupiter is passing through the perihelion of its orbit, bringing it closer to Earth.[132] Near opposition, Jupiter will appear to go into retrograde motion for a period of about 121 days, moving backward through an angle of 9.9° before returning to prograde movement.[133]

Because the orbit of Jupiter is outside that of Earth, the phase angle of Jupiter as viewed from Earth is always less than 11.5°; thus, Jupiter always appears nearly fully illuminated when viewed through

Earth-based telescopes. It was only during spacecraft missions to Jupiter that crescent views of the planet were obtained.[134] A small telescope will usually show Jupiter's four Galilean moons and the prominent cloud belts across Jupiter's atmosphere. A larger telescope with an aperture of 4–6 in (10.16–15.24 cm) will show Jupiter's Great Red Spot when it faces Earth.[135][136]

History

Pre-telescopic research

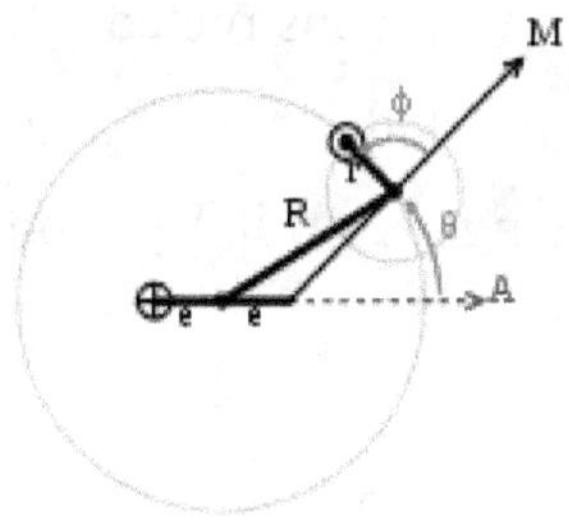

Model in the Almagest of the longitudinal motion of Jupiter (☉) relative to Earth (□)

Observation of Jupiter dates back to at least the Babylonian astronomers of the 7th or 8th century BC.[137] The ancient Chinese knew Jupiter as the "*Suì* Star" (*Suìxīng* 歲星) and established their cycle of 12 earthly branches based on the approximate number of years it takes Jupiter to rotate around the Sun; the Chinese language still uses its name (simplified as 歲) when referring to years of age.

By the 4th century BC, these observations had developed into the Chinese zodiac,[138] and each year became associated with a Tai Sui star and god controlling the region of the heavens opposite Jupiter's position in the night sky. These beliefs survive in some

Taoistreligious practices and in the East Asian zodiac's twelve animals. The Chinese historian Xi Zezong has claimed that Gan De, an ancient Chinese astronomer,[139] reported a small star "in alliance" with the planet,[140] which may indicate a sighting of one of Jupiter's moonswith the unaided eye. If true, this would predate Galileo's discovery by nearly two millennia.[141][142]

A 2016 paper reports that trapezoidal rule was used by Babylonians before 50 BCE for integrating the velocity of Jupiter along the ecliptic.[143] In his 2nd century work the Almagest, the Hellenistic astronomer Claudius Ptolemaeus constructed a geocentricplanetary model based on deferents and epicycles to explain Jupiter's motion relative to Earth, giving its orbital period around Earth as 4332.38 days, or 11.86 years.[144]

Ground-based telescope research

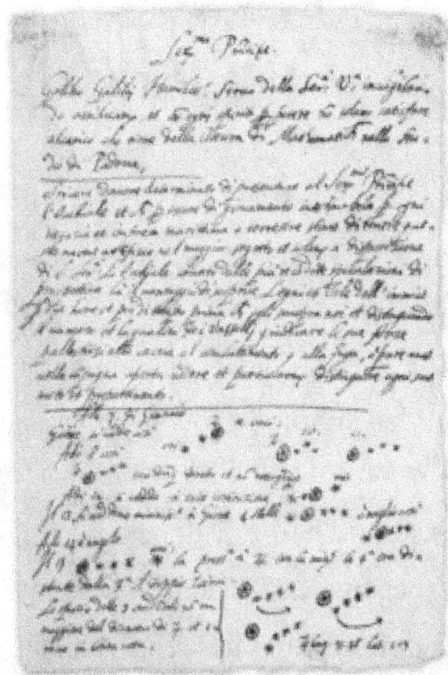

Galileo's original observation note of Jupiter moons

In 1610, Italian polymath Galileo Galileidiscovered the four largest moons of Jupiter (now known as the Galilean moons) using a telescope. This is thought to be the first telescopic observation of moons other than Earth's. Just one day after Galileo, Simon Marius independently discovered moons around Jupiter, though he did not publish his discovery in a book until 1614.[145] It was Marius's names for the major moons, however, that stuck: Io, Europa, Ganymede, and Callisto. The discovery was a major point in favour of Copernicus' heliocentric theory of the motions of the planets; Galileo's outspoken support of the Copernican theory led to him being tried and condemned by the Inquisition.[146]

During the 1660s, Giovanni Cassini used a new telescope to discover spots and colourful bands in Jupiter's atmosphere, observe that the planet appeared oblate, and estimate its rotation period.[147] In 1692, Cassini noticed that the atmosphere undergoes differential rotation.[148]

The Great Red Spot may have been observed as early as 1664 by Robert Hooke and in 1665 by Cassini, although this is disputed. The pharmacist Heinrich Schwabe produced the earliest known drawing to show details of the Great Red Spot in 1831.[149] The Red Spot was reportedly lost from sight on several occasions between 1665 and 1708 before becoming quite conspicuous in 1878.[150] It was recorded as fading again in 1883 and at the start of the 20th century.[151]

Both Giovanni Borelli and Cassini made careful tables of the motions of Jupiter's moons, which allowed predictions of when the moons would pass before or behind the planet. By the 1670s, Cassini observed that when Jupiter was on the opposite side of the Sun from Earth, these events would occur about 17 minutes later than expected. Ole Rømer deduced that light does not travel instantaneously (a conclusion that Cassini had earlier rejected),[49] and this timing discrepancy was used to estimate the speed of light.[152][153]

In 1892, E. E. Barnard observed a fifth satellite of Jupiter with the 36-inch (910 mm) refractor at Lick Observatory in California. This moon was later named Amalthea.[154] It was the last planetary moon to be discovered directly by a visual observer through a telescope.[155] An additional eight satellites were discovered before the flyby of the Voyager 1 probe in 1979.[e]

Jupiter viewed in infrared by JWST

(July 14, 2022)

In 1932, Rupert Wildt identified absorption bands of ammonia and methane in the spectra of Jupiter.[156] Three long-lived anticyclonic features called "white ovals" were observed in 1938. For several decades they remained as separate features in the atmosphere, sometimes approaching each other but never merging. Finally, two of the ovals merged in 1998, then absorbed the third in 2000, becoming Oval BA.[157]

Space-based telescope research

On July 14, 2022, NASA presented images of Jupiter and related areas captured, for the first time, and including infrared views, by the James Webb Space Telescope (JWST).[158]

Radiotelescope research

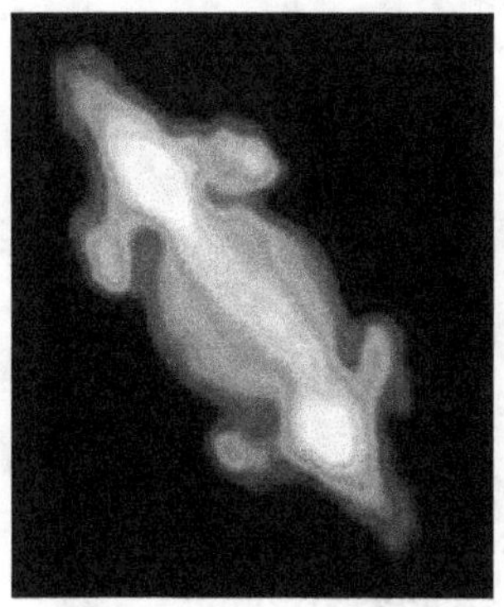

Image of Jupiter and its radiation belts in radio

In 1955, Bernard Burke and Kenneth Franklindiscovered that Jupiter emits bursts of radio waves at a frequency of 22.2 MHz.[64]:36 The period of these bursts matched the rotation of the planet, and they used this information to determine a more precise value for Jupiter's rotation rate. Radio bursts from Jupiter were found to come in two forms: long bursts (or L-bursts) lasting up to several seconds, and short bursts (or S-bursts) lasting less than a hundredth of a second.[159]

Scientists have discovered three forms of radio signals transmitted from Jupiter:

- Decametric radio bursts (with a wavelength of tens of metres) vary with the rotation of Jupiter, and are influenced by the interaction of Io with Jupiter's magnetic field.[160]

- Decimetric radio emission (with wavelengths measured in centimetres) was first observed by Frank Drake and Hein Hvatum in 1959.[64]:36 The origin of this signal is a torus-shaped belt around Jupiter's equator, which generates cyclotron radiation from electrons that are accelerated in Jupiter's magnetic field.[161]

- Thermal radiation is produced by heat in the atmosphere of Jupiter.[64]:43

Exploration

Main article: Exploration of Jupiter

Jupiter has been visited by automated spacecraft since 1973, when the space probe Pioneer 10 passed close enough to Jupiter to send back revelations about its properties and phenomena.[162][163] Missions to Jupiter are accomplished at a cost in energy, which is described by the net change in velocity of the spacecraft, or delta-v. Entering a Hohmann transfer orbit from Earth to Jupiter from low Earth orbit requires a delta-v of 6.3 km/s,[164]which is comparable to the 9.7 km/s delta-v needed to reach low Earth orbit.[165] Gravity assists through planetary flybys can be used to reduce the energy required to reach Jupiter.[166]

Flyby missions

SpacecraftClosest

approachDistance

Pioneer 10

December 3, 1973

130,000 km

<u>Pioneer 11</u>

December 4, 1974

34,000 km

<u>Voyager 1</u>

March 5, 1979

349,000 km

<u>Voyager 2</u>

July 9, 1979

570,000 km

<u>Ulysses</u>

February 8, 1992[167]

408,894 km

February 4, 2004[167]

120,000,000 km

<u>Cassini</u>

December 30, 2000

10,000,000 km

<u>New Horizons</u>

February 28, 2007

2,304,535 km

Beginning in 1973, several spacecraft have performed planetary flyby manoeuvres that brought them within observation range of Jupiter. The <u>Pioneer</u> missions obtained the first close-up images of Jupiter's atmosphere and several of its moons. They discovered that the radiation fields near the planet were much stronger than expected, but both spacecraft managed to survive in that environment. The trajectories of these spacecraft were used to refine the mass estimates of the Jovian system. <u>Radio occultations</u> by the planet resulted in better measurements of Jupiter's diameter and the amount of polar flattening.[55]:47 [168]

Six years later, the Voyager missions vastly improved the understanding of the Galilean moons and discovered Jupiter's rings. They also confirmed that the Great Red Spot was anticyclonic. Comparison of images showed that the Spot had changed hue since the Pioneer missions, turning from orange to dark brown. A torus of ionized atoms was discovered along Io's orbital path, which were found to come from erupting volcanoes on the moon's surface. As the spacecraft passed behind the planet, it observed flashes of lightning in the night side atmosphere.[55]:87 [169]

The next mission to encounter Jupiter was the Ulysses solar probe. In February 1992, it performed a flyby manoeuvre to attain a polar orbit around the Sun. During this pass, the spacecraft studied Jupiter's magnetosphere, although it had no cameras to photograph the planet. The spacecraft passed by Jupiter six years later, this time at a much greater distance.[167]

In 2000, the *Cassini* probe flew by Jupiter on its way to Saturn, and provided higher-resolution images.[170]

The New Horizons probe flew by Jupiter in 2007 for a gravity assist en route to Pluto.[171]The probe's cameras measured plasma output from volcanoes on Io and studied all four Galilean moons in detail.[172]

Galileo mission

Main article: Galileo (spacecraft)

Galileo in preparation for mating with the rocket, 2000

The first spacecraft to orbit Jupiter was the Galileo mission, which reached the planet on December 7, 1995.[60] It remained in orbit for over seven years, conducting multiple flybys of all the Galilean moons and Amalthea. The spacecraft also witnessed the impact of Comet Shoemaker–Levy 9 when it collided with Jupiter in 1994. Some of the goals for the mission were thwarted due to a malfunction in *Galileo*'s high-gain antenna.[173]

A 340-kilogram titanium atmospheric probe was released from the spacecraft in July 1995, entering Jupiter's atmosphere on December 7.[60] It parachuted through 150 km (93 mi) of the atmosphere at a speed of about 2,575 km/h (1600 mph)[60] and collected data for 57.6 minutes until the spacecraft was destroyed.[174] The *Galileo* orbiter itself experienced a more rapid version of the same fate when it was deliberately steered into the planet on September 21, 2003. NASA destroyed the spacecraft in order to avoid any possibility of the spacecraft crashing into and possibly contaminating the moon Europa, which may harbour life.[173]

Data from this mission revealed that hydrogen composes up to 90% of Jupiter's atmosphere.[60] The recorded temperature was more than 300 °C (570 °F) and the windspeed measured more than 644 km/h (>400 mph) before the probes vaporized.[60]

Juno mission

Main article: Juno (spacecraft)

Juno preparing for testing in a rotation stand, 2011

NASA's *Juno* mission arrived at Jupiter on July 4, 2016 with the goal of studying the planet in detail from a polar orbit. The spacecraft was originally intended to orbit Jupiter thirty-seven times over a period of twenty months.[175][19][176] During the mission, the spacecraft will be exposed to high levels of radiation from Jupiter's magnetosphere, which may cause future failure of certain instruments.[177] On August 27, 2016, the spacecraft completed its first fly-by of Jupiter and sent back the first ever images of Jupiter's north pole.[178]

Juno completed 12 orbits before the end of its budgeted mission plan, ending July 2018.[179]In June of that year, NASA extended the mission operations plan to July 2021, and in January of that year the mission was extended to September 2025 with four lunar flybys: one of Ganymede, one of Europa, and two of Io.[180][181] When *Juno* reaches the end of the mission, it will perform a controlled deorbit and disintegrate into Jupiter's atmosphere. This will avoid the risk of collision with Jupiter's moons.[182][183]

Cancelled missions and future plans

There is great interest in missions to study Jupiter's larger icy moons, which may have subsurface liquid oceans. Funding difficulties have delayed progress, causing NASA's JIMO(*Jupiter Icy Moons Orbiter*) to be cancelled in 2005.[184] A subsequent proposal was developed for a joint NASA/ESA mission called EJSM/Laplace, with a provisional launch date around 2020. EJSM/Laplace would have consisted of the NASA-led Jupiter Europa Orbiter and the ESA-led Jupiter Ganymede Orbiter.[185] However, the ESA formally ended the partnership in April 2011, citing budget issues at NASA and the consequences on the mission timetable. Instead, ESA planned to go ahead with a European-only mission to compete in its L1 Cosmic Vision selection.[186]These plans have been realized as the European Space Agency's Jupiter Icy

Moon Explorer (JUICE), due to launch in 2023,[187]followed by NASA's Europa Clipper mission, scheduled for launch in 2024.[188]

Other proposed missions include the Chinese National Space Administration's Gan Demission which aims to launch an orbiter to the Jovian system and possibly Callisto around 2035,[189] and CNSA's Interstellar Express[190]and NASA's Interstellar,[191] which would both use Jupiter's gravity to help them reach the edges of the heliosphere.

Moons

Main article: Moons of Jupiter

See also: Timeline of discovery of Solar System planets and their moons *and* Satellite system (astronomy)

Jupiter has 80 known natural satellites.[6][192]Of these, 60 are less than 10 km in diameter.[193] The four largest moons are Io, Europa, Ganymede, and Callisto, collectively known as the "Galilean moons", and are visible from Earth with binoculars on a clear night.[194]

Galilean moons

Main article: Galilean moons

The moons discovered by Galileo—Io, Europa, Ganymede, and Callisto—are among the largest in the Solar System. The orbits of Io, Europa, and Ganymede form a pattern known as a Laplace resonance; for every four orbits that Io makes around Jupiter, Europa makes exactly two orbits and Ganymede makes exactly one. This resonance causes the gravitational effects of the three large moons to distort their orbits into elliptical shapes, because each moon receives an extra tug from its neighbours at the same point in every orbit it makes. The tidal force from Jupiter, on the other hand, works to circularise their orbits.[195]

The eccentricity of their orbits causes regular flexing of the three moons' shapes, with Jupiter's gravity stretching them out as they approach it and allowing them to spring back to more spherical shapes as they swing away. The friction created by this tidal flexing generates heat in the interior of the moons.[196]This is seen most dramatically in the volcanic activity of Io (which is subject to the strongest tidal forces),[196] and to a lesser degree in the geological youth of Europa's surface, which indicates recent resurfacing of the moon's exterior.[197]

The Galilean moons, as a percent of the Earth's Moon

NameIPADiameterMassOrbital radiusOrbital period

km%kg%km%days%

Io

/ˈaɪ.oʊ/

3,643

105

8.9×10^{22}

120

421,700

110

1.77

7

Europa

/jʊˈroʊpə/

3,122

90

$4.8×10^{22}$

65

671,034

175

3.55

13

Ganymede

/ˈɡænimiːd/

5,262

150

$14.8×10^{22}$

200

1,070,412

280

7.15

26

Callisto

/kəˈlɪstoʊ/

4,821

140

$10.8×10^{22}$

150

1,882,709

490

16.69

61

The Galilean moons Io, Europa, Ganymede, and Callisto (in order of increasing distance from Jupiter)

Classification

Jupiter's moons were traditionally classified into four groups of four, based on their similar orbital elements.[198] This picture has been complicated by the discovery of numerous small outer moons since 1999. Jupiter's moons are currently divided into several different groups, although there are several moons which are not part of any group.[199]

The eight innermost regular moons, which have nearly circular orbits near the plane of Jupiter's equator, are thought to have formed alongside Jupiter, whilst the remainder are irregular moons and are thought to be captured asteroids or fragments of captured asteroids. The irregular moons within each group may have a common origin, perhaps as a larger moon or captured body that broke up.[200][201]

Regular moons

Inner group

The inner group of four small moons all have diameters of less than 200 km, orbit at radii less than 200,000 km, and have orbital inclinations of less than half a degree.

These four moons, discovered by Galileo Galilei and by Simon Marius in parallel, orbit between 400,000 and 2,000,000 km, and are some of the largest moons in the Solar System.

Irregular moons

Himalia group

A tightly clustered group of moons with orbits around 11,000,000–12,000,000 km from Jupiter.[203]

Ananke group

This retrograde orbitgroup has rather indistinct borders, averaging 21,276,000 km from Jupiter with an average inclination of 149 degrees.[201]

Carme group

A fairly distinct retrograde group that averages 23,404,000 km from Jupiter with an average inclination of 165 degrees.[201]

Pasiphae group

A dispersed and only vaguely distinct retrograde group that covers all the outermost moons.[204]

Interaction with the Solar System

As the most massive of the eight planets, the gravitational influence of Jupiter has helped shape the Solar System. With the exception of Mercury, the orbits of the system's planets lie closer to Jupiter's orbital plane than the Sun's equatorial plane. The Kirkwood gaps in the asteroid belt are mostly caused by Jupiter,[205]and the planet may have been responsible for the Late Heavy Bombardment in the inner Solar System's history.[206]

In addition to its moons, Jupiter's gravitational field controls numerous asteroids that have settled around the Lagrangian points that precede and follow the planet in its orbit around the Sun. These are known as the Trojan asteroids, and are divided into Greek and Trojan "camps" to honour the Iliad. The first of these, 588 Achilles, was discovered by Max Wolf in 1906; since then more than two thousand have been discovered.[207] The largest is 624 Hektor.[208]

The Jupiter family is defined as comets that have a semi-major axis smaller than Jupiter's; most short-period comets belong to this group. Members of the Jupiter family are thought to form in the Kuiper belt outside the orbit of Neptune. During close encounters with Jupiter, they are perturbed into orbits with a smaller period, which then becomes circularised by regular gravitational interaction with the Sun and Jupiter.[209]

Impacts

Main article: Impact events on Jupiter

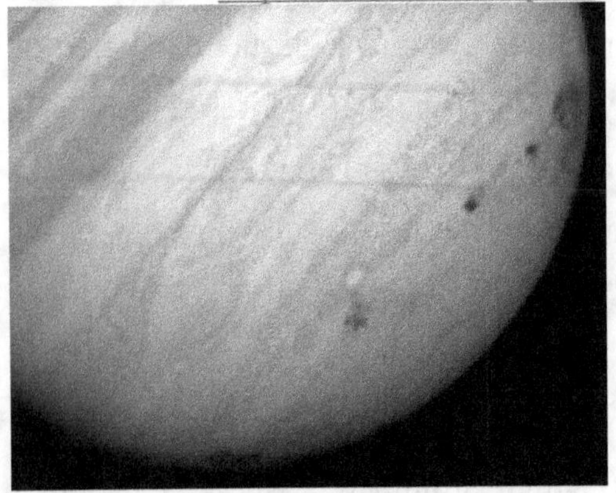

Brown spots mark Comet Shoemaker–Levy 9's impact sites on Jupiter

Jupiter has been called the Solar System's vacuum cleaner[210] because of its immense gravity well and location near the inner Solar System. There are more impacts on Jupiter, such as comets, than on any other planet in the Solar System.[211] For example, Jupiter experiences about 200 times more asteroid and comet impacts than Earth.[60] In the past, scientists believed that Jupiter partially shielded the inner system from cometary bombardment.[60] However, computer simulations in 2008 suggest that Jupiter does not cause a net decrease in the number of comets that pass through the inner Solar System, as its gravity perturbs their orbits inward roughly as often as it accretes or ejects them.[212] This topic remains controversial among scientists, as some think it draws comets towards Earth from the Kuiper belt, while others believes that Jupiter protects Earth from the Oort cloud.[213]

In July 1994, the Comet Shoemaker–Levy 9comet collided with Jupiter.[214][215] The impacts were closely observed by observatories around the world, including the Hubble Space Telescope and Galileospacecraft.[216][217][218][219] The event was widely covered by the media.[220]

Surveys of early astronomical records and drawings produced eight examples of potential impact observations between 1664 and 1839. However, a 1997 review determined that these observations had little or no possibility of being the results of impacts. Further investigation by this team revealed a dark surface feature discovered by astronomer Giovanni Cassini in 1690 may have been an impact scar.[221]

In culture

See also: Jupiter in fiction *and* Planets in astrology § Jupiter

Jupiter, woodcut from a 1550 edition of Guido Bonatti's *Liber Astronomiae*

The planet Jupiter has been known since ancient times. It is visible to the naked eye in the night sky and can occasionally be seen in the daytime when the Sun is low.[222] To the Babylonians, this planet represented their god Marduk,[223] chief of their pantheon from the Hammurabi period.[224] They used Jupiter's roughly 12-year orbit along the ecliptic to define the constellations of their zodiac.[223]

The mythical Greek name for this planet is Zeus (Ζεύς), also referred to as *Dias* (Δίας), the planetary name of which is retained in modern Greek.[225] The ancient Greeks knew the planet as Phaethon (Φαέθων), meaning "shining one" or "blazing star".[226][227] The Greek myths of Zeus from the Homeric period showed particular similarities to certain Near-Easterngods, including the Semitic El and Baal, the Sumerian Enlil, and the Babylonian god Marduk.[228] The association between the planet and the Greek deity Zeus was drawn from Near Eastern influences and was fully established by the fourth century BCE, as documented in the Epinomis of Plato and his contemporaries.[229]

The god Jupiter is the Roman counterpart of Zeus, and he is the principal god of Roman mythology. The Romans originally called Jupiter the "star of Jupiter" (*Iuppiter Stella*)," as they believed it to be sacred to its namesake god. This name comes from the Proto-Indo-European vocative compound *Dyēu-pəter*(nominative: *Dyēus-pətēr*, meaning "Father Sky-God", or "Father Day-God").[230] As the supreme god of the Roman pantheon, Jupiter was the god of thunder, lightning, and storms, and appropriately called the god of light and sky.

In Vedic astrology, Hindu astrologers named the planet after Brihaspati, the religious teacher of the gods, and often called it "Guru", which means the "Teacher".[231][232] In Central Asian Turkic myths, Jupiter is called *Erendiz* or *Erentüz*, from *eren* (of uncertain meaning) and *yultuz* ("star"). The Turks calculated the period of the orbit of Jupiter as 11 years and 300 days. They believed that some social and natural events connected to Erentüz's movements on the sky.[233] The Chinese, Vietnamese, Koreans, and Japanese called it the "wood star" (Chinese: 木星; pinyin: mùxīng), based on the Chinese Five Elements.[234][235][236] In China it became known as the "Year-star" (Sui-sing) as Chinese astronomers noted that it jumped one zodiac constellation each year (with corrections). In some ancient Chinese writings the years were named, at least in principle, in correlation with the Jovian zodiacal signs.[237]

Gallery

-

- Infrared view of Jupiter, imaged by the Gemini North telescope in Hawai'i on January 11, 2017

-
- Jupiter imaged in visible light by the Hubble Space Telescope on January 11, 2017

-

- Ultraviolet view of Jupiter, imaged by Hubble on January 11, 2017[238]

-

- This image of Jupiter and Europa, taken by Hubble on 25 August 2020, was captured when the planet was 653 million kilometres from Earth.[239]

See also

- solar system portal

- outer space portal

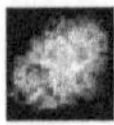

- astronomy portal

- Eccentric Jupiter – Jovian planet that orbits its star in an eccentric orbit

- <u>Hot Jupiter</u> – Class of high mass planets orbiting close to a star
- <u>Jovian–Plutonian gravitational effect</u> – Astronomical hoax
- <u>List of gravitationally rounded objects of the Solar System</u>
- <u>March 17, 2016 collision with Jupiter</u>
- <u>Outline of Jupiter</u> – Overview of and topical guide to Jupiter
- <u>Super-Jupiter</u> – Class of planets with more mass than Jupiter

Notes

1. ^ On the left side of the photo is <u>Europa</u>, and the red spot at the planet's center is called the <u>Great Red Spot</u>
2. ^ <u>a</u> <u>b</u> <u>c</u> <u>d</u> <u>e</u> <u>f</u> Refers to the level of 1 bar atmospheric pressure
3. ^ Based on the volume within the level of 1 bar atmospheric pressure
4. ^ See for example: <u>"IAUC 2844: Jupiter; 1975h"</u>. International Astronomical Union. October 1, 1975. Retrieved October 24, 2010. That particular word has been in use since at least 1966. See: <u>"Query Results from the Astronomy Database"</u>. Smithsonian/NASA. Retrieved July 29, 2007.
5. ^ See <u>Moons of Jupiter</u> for details and cites

References

1. ^ Simpson, J. A.; Weiner, E. S. C. (1989). <u>"Jupiter"</u>. *Oxford English Dictionary*. Vol. 8 (2nd ed.). Clarendon Press. <u>ISBN 978-0-19-861220-9</u>.
2. ^ <u>a</u> <u>b</u> Seligman, Courtney. <u>"Rotation Period and Day Length"</u>. Retrieved August 13, 2009.

3. ^ a b c d Simon, J. L.; Bretagnon, P.; Chapront, J.; Chapront-Touzé, M.; Francou, G.; Laskar, J. (February 1994). "Numerical expressions for precession formulae and mean elements for the Moon and planets". *Astronomy and Astrophysics*. **282** (2): 663–683. Bibcode:1994A&A...282..663S.

4. ^ Souami, D.; Souchay, J. (July 2012). "The solar system's invariable plane". *Astronomy & Astrophysics*. **543**: 11. Bibcode:2012A&A...543A.133S. doi:10.1051/0004-6361/201219011. A133.

5. ^ "HORIZONS Planet-center Batch call for January 2023 Perihelion". *ssd.jpl.nasa.gov*(Perihelion for Jupiter's planet-centre (599) occurs on 2023-Jan-21 at 4.9510113au during a rdot flip from negative to positive). NASA/JPL. Retrieved September 7, 2021.

6. ^ a b c Kindy, David. "Amateur Astronomer Discovers New Moon Orbiting Jupiter". *Smithsonian Magazine*. Retrieved March 8,2022.

7. ^ a b c d e f g Williams, David R. (December 23, 2021). "Jupiter Fact Sheet". NASA. Retrieved October 13, 2017.

8. ^ "Astrodynamic Constants". JPL Solar System Dynamics. February 27, 2009. Retrieved August 8, 2007.

9. ^ Ni, D. (2018). "Empirical models of Jupiter's interior from Juno data". *Astronomy & Astrophysics*. **613**: A32. Bibcode:2018A&A...613A..32N. doi:10.1051/0004-6361/201732183.

10. ^ Li, Liming; Jiang, X.; West, R. A.; Gierasch, P. J.; Perez-Hoyos, S.; Sanchez-Lavega, A.; Fletcher, L. N.; Fortney, J.

J.; Knowles, B.; Porco, C. C.; Baines, K. H.; Fry, P. M.; Mallama, A.; Achterberg, R. K.; Simon, A. A.; Nixon, C. A.; Orton, G. S.; Dyudina, U. A.; Ewald, S. P.; Schmude, R. W. (2018). "Less absorbed solar energy and more internal heat for Jupiter". *Nature Communications*. **9** (1): 3709. Bibcode:2018NatCo...9.3709L. doi:10.1038/s41467-018-06107-2. PMC 6137063. PMID 30213944.

11. ^ Mallama, Anthony; Krobusek, Bruce; Pavlov, Hristo (2017). "Comprehensive wide-band magnitudes and albedos for the planets, with applications to exo-planets and Planet Nine". *Icarus*. **282**: 19–33. arXiv:1609.05048. Bibcode:2017Icar..282...19M. doi:10.1016/j.icarus.2016.09.023. S2CID 119307693.

12. ^ a b c d Mallama, A.; Hilton, J. L. (2018). "Computing Apparent Planetary Magnitudes for The Astronomical Almanac". *Astronomy and Computing*. **25**: 10–24. arXiv:1808.01973. Bibcode:2018A&C....25...10M. doi:10.1016/j.ascom.2018.08.002. S2CID 69912809.

13. ^ Seidelmann, P. Kenneth; Archinal, Brent A.; A'Hearn, Michael F.; Conrad, Albert R.; Consolmagno, Guy J.; Hestroffer, Daniel; Hilton, James L.; Krasinsky, Georgij A.; Neumann, Gregory A.; Oberst, Jürgen; Stooke, Philip J.; Tedesco, Edward F.; Tholen, David J.; Thomas, Peter C.; Williams, Iwan P. (2007). "Report of the IAU/IAG Working Group on cartographic coordinates and rotational elements: 2006". *Celestial Mechanics and Dynamical Astronomy*. **98** (3): 155–180. Bibcode:2007CeMDA..98..155S. doi:10.1007/s10569-007-9072-y.

14. ^ de Pater, Imke; Lissauer, Jack J. (2015). *Planetary Sciences* (2nd updated ed.). New York: Cambridge University Press. p. 250. ISBN 978-0-521-85371-2.

15. ^ Bjoraker, G. L.; Wong, M. H.; de Pater, I.; Ádámkovics, M. (September 2015). "Jupiter's Deep Cloud Structure Revealed Using Keck Observations of Spectrally Resolved Line Shapes". *The Astrophysical Journal*. **810** (2): 10. arXiv:1508.04795. Bibcode:2015ApJ...810..122B. doi:10.1088/0004-637X/810/2/122. S2CID 55592285. 122.

16. ^ Saumon, D.; Guillot, T. (2004). "Shock Compression of Deuterium and the Interiors of Jupiter and Saturn". *The Astrophysical Journal*. **609** (2): 1170–1180. arXiv:astro-ph/0403393. Bibcode:2004ApJ...609.1170S. doi:10.1086/421257. S2CID 119325899.

17. ^ "In Depth | Pioneer 10". *NASA Solar System Exploration*. Retrieved February 9, 2020. Pioneer 10, the first NASA mission to the outer planets, garnered a series of firsts perhaps unmatched by any other robotic spacecraft in the space era: the first vehicle placed on a trajectory to escape the solar system into interstellar space; the first spacecraft to fly beyond Mars; the first to fly through the asteroid belt; the first to fly past Jupiter; and the first to use all-nuclear electrical power

18. ^ "Exploration | Jupiter". *NASA Solar System Exploration*. Retrieved February 9, 2020.

19. ^ a b c Chang, Kenneth (July 5, 2016). "NASA's Juno Spacecraft Enters Jupiter's Orbit". The New York Times. Retrieved July 5, 2016.

20. ^ Chang, Kenneth (June 30, 2016). "All Eyes (and Ears) on Jupiter". The New York Times. Retrieved July 1, 2016.

21. ^ Chang, Kenneth (June 14, 2021). "Mushballs and a Great Blue Spot: What Lies Beneath Jupiter's Pretty Clouds – NASA's Juno probe is beginning an extended mission that may not have been possible if it hadn't experienced engine trouble when it first arrived at the giant planet". The New York Times. Archived from the original on December 28, 2021. Retrieved June 16, 2021.

22. ^ "Naming of Astronomical Objects". International Astronomical Union. Retrieved March 23, 2022.

23. ^ Jones, Alexander (1999). *Astronomical papyri from Oxyrhynchus*. pp. 62–63. ISBN 9780871692337. It is now possible to trace the medieval symbols for at least four of the five planets to forms that occur in some of the latest papyrus horoscopes ([P.Oxy.] 4272, 4274, 4275 […]). That for Jupiter is an obvious monogram derived from the initial letter of the Greek name.

24. ^ Maunder, A. S. D. (August 1934). "The origin of the symbols of the planets". *The Observatory*. **57**: 238–247. Bibcode:1934Obs....57..238M.

25. ^ Harper, Douglas. *Jove*. *Online Etymology Dictionary*. Retrieved March 22, 2022.

26. ^ Falk, Michael (June 1999), "Astronomical Names for the Days of the Week", *Journal of the Royal Astronomical Society of Canada*, **93**: 122–133, Bibcode:1999JRASC..93..122F, archived from the original on February 25, 2021, retrieved November 18, 2020

27. ^ Falk, Michael; Koresko, Christopher (2004). "Astronomical Names for the Days of the Week". *Journal of the Royal Astronomical Society of Canada*. **93**: 122–133. arXiv:astro-ph/0307398. Bibcode:1999JRASC..93..122F. doi:10.1016/j.newast.2003.07.002. S2CID 118954190.

28. ^ "Jovial". *Dictionary.com*. Retrieved July 29,2007.

29. ^ a b Kruijer, Thomas S.; Burkhardt, Christoph; Budde, Gerrit; Kleine, Thorsten (June 2017). "Age of Jupiter inferred from the distinct genetics and formation times of meteorites". *Proceedings of the National Academy of Sciences*. **114** (26): 6712–6716. Bibcode:2017PNAS..114.6712K. doi:10.1073/pnas.1704461114. PMC 5495263. PMID 28607079.

30. ^ a b Bosman, A. D.; Cridland, A. J.; Miguel, Y. (December 2019). "Jupiter formed as a pebble pile around the N2 ice line". *Astronomy & Astrophysics*. **632**: 5. arXiv:1911.11154. Bibcode:2019A&A...632L..11B. doi:10.1051/0004-6361/201936827. S2CID 208291392. L11.

31. ^ a b Walsh, K. J.; Morbidelli, A.; Raymond, S. N.; O'Brien, D. P.; Mandell, A. M. (2011). "A low mass for Mars from Jupiter's early gas-driven migration". Nature. **475** (7355): 206–209. arXiv:1201.5177. Bibcode:2011Natur.475..206W. doi:10.1038/nature10201. PMID 21642961. S2CID 4431823.

32. ^ Batygin, Konstantin (2015). "Jupiter's decisive role in the inner Solar System's early evolution". *Proceedings of the National Academy of Sciences*. **112** (14): 4214–4217. arXiv:1503.06945. Bibcode:2015PNAS..112.4214B.

doi:10.1073/pnas.1423252112. PMC 4394287. PMID 25831540.

33. ^ Haisch Jr., K. E.; Lada, E. A.; Lada, C. J. (2001). "Disc Frequencies and Lifetimes in Young Clusters". *The Astrophysical Journal*. **553** (2): 153–156. arXiv:astro-ph/0104347. Bibcode:2001ApJ...553L.153H. doi:10.1086/320685. S2CID 16480998.

34. ^ Fazekas, Andrew (March 24, 2015). "Observe: Jupiter, Wrecking Ball of Early Solar System". *National Geographic*. Archived from the original on March 14, 2017. Retrieved April 18,2021.

35. ^ Zube, N.; Nimmo, F.; Fischer, R.; Jacobson, S. (2019). "Constraints on terrestrial planet formation timescales and equilibration processes in the Grand Tack scenario from Hf-W isotopic evolution". *Earth and Planetary Science Letters*. **522** (1): 210–218. arXiv:1910.00645. Bibcode:2019E&PSL.522..210Z. doi:10.1016/j.epsl.2019.07.001. PMC 7339907. PMID 32636530. S2CID 199100280.

36. ^ D'Angelo, G.; Marzari, F. (2012). "Outward Migration of Jupiter and Saturn in Evolved Gaseous Disks". *The Astrophysical Journal*. **757**(1): 50 (23 pp.). arXiv:1207.2737. Bibcode:2012ApJ...757...50D. doi:10.1088/0004-637X/757/1/50. S2CID 118587166.

37. ^ D'Angelo, G.; Weidenschilling, S. J.; Lissauer, J. J.; Bodenheimer, P. (2021). "Growth of Jupiter: Formation in disks of gas and solids and evolution to the present epoch". *Icarus*. **355**: 114087. arXiv:2009.05575.

Bibcode:2021Icar..35514087D.
doi:10.1016/j.icarus.2020.114087. S2CID 221654962.

38. ^ a b Pirani, S.; Johansen, A.; Bitsch, B.; Mustill, A.J.; Turrini, D. (March 2019). "Consequences of planetary migration on the minor bodies of the early solar system". *Astronomy & Astrophysics*. **623**: A169. arXiv:1902.04591. Bibcode:2019A&A...623A.169P. doi:10.1051/0004-6361/201833713.

39. ^ a b "Jupiter's Unknown Journey Revealed". *ScienceDaily*. Lund University. March 22, 2019. Retrieved March 25, 2019.

40. ^ Öberg, K.I.; Wordsworth, R. (2019). "Jupiter's Composition Suggests its Core Assembled Exterior to the N_{2} Snowline". *The Astronomical Journal*. **158** (5). arXiv:1909.11246. doi:10.3847/1538-3881/ab46a8. S2CID 202749962.

41. ^ Öberg, K.I.; Wordsworth, R. (2020). "Erratum: "Jupiter's Composition Suggests Its Core Assembled Exterior to the N2 Snowline"". *The Astronomical Journal*. **159** (2): 78. doi:10.3847/1538-3881/ab6172. S2CID 214576608.

42. ^ Denecke, Edward J. (January 7, 2020). *Regents Exams and Answers: Earth Science—Physical Setting 2020*. Barrons Educational Series. p. 419. ISBN 978-1-5062-5399-2.

43. ^ Swarbrick, James (2013). *Encyclopedia of Pharmaceutical Technology*. Vol. 6. CRC Press. p. 3601. ISBN 9781439808238. Syrup USP (1.31 g/cm^3)

44. ^ Allen, Clabon Walter; Cox, Arthur N. (2000). *Allen's Astrophysical Quantities*. Springer. pp. 295–296. ISBN 978-0-387-98746-0.

45. ^ Polyanin, Andrei D.; Chernoutsan, Alexei (October 18, 2010). *A Concise Handbook of Mathematics, Physics, and Engineering Sciences*. CRC Press. p. 1041. ISBN 978-1-4398-0640-1.

46. ^ Guillot, Tristan; Gautier, Daniel; Hubbard, William B (December 1997). "NOTE: New Constraints on the Composition of Jupiter from Galileo Measurements and Interior Models". *Icarus*. **130** (2): 534–539. arXiv:astro-ph/9707210. Bibcode:1997Icar..130..534G. doi:10.1006/icar.1997.5812. S2CID 5466469.

47. ^ Kim, S. J.; Caldwell, J.; Rivolo, A. R.; Wagner, R. (1985). "Infrared Polar Brightening on Jupiter III. Spectrometry from the Voyager 1 IRIS Experiment". *Icarus*. **64** (2): 233–248. Bibcode:1985Icar...64..233K. doi:10.1016/0019-1035(85)90201-5.

48. ^ Gautier, D.; Conrath, B.; Flasar, M.; Hanel, R.; Kunde, V.; Chedin, A.; Scott N. (1981). "The helium abundance of Jupiter from Voyager". *Journal of Geophysical Research*. **86** (A10): 8713–8720. Bibcode:1981JGR....86.8713G. doi:10.1029/JA086iA10p08713. hdl:2060/19810016480. S2CID 122314894.

49. ^ a b Kunde, V. G.; et al. (September 10, 2004). "Jupiter's Atmospheric Composition from the Cassini Thermal Infrared Spectroscopy Experiment". *Science*. **305** (5690): 1582–1586. Bibcode:2004Sci...305.1582K. doi:10.1126/science.1100240. PMID 15319491. S2CID 45296656. Retrieved April 4, 2007.

50. ^ Niemann, H. B.; Atreya, S. K.; Carignan, G. R.; Donahue, T. M.; Haberman, J. A.; Harpold, D. N.; Hartle, R. E.; Hunten, D. M.; Kasprzak, W. T.; Mahaffy, P. R.; Owen, T. C.; Spencer, N. W.; Way, S. H. (1996). "The Galileo Probe Mass Spectrometer: Composition of Jupiter's Atmosphere". *Science*. **272** (5263): 846–849. Bibcode:1996Sci...272..846N. doi:10.1126/science.272.5263.846. PMID 8629016. S2CID 3242002.

51. ^ a b von Zahn, U.; Hunten, D. M.; Lehmacher, G. (1998). "Helium in Jupiter's atmosphere: Results from the Galileo probe Helium Interferometer Experiment". *Journal of Geophysical Research*. **103** (E10): 22815–22829. Bibcode:1998JGR...10322815V. doi:10.1029/98JE00695.

52. ^ a b Stevenson, David J. (May 2020). "Jupiter's Interior as Revealed by Juno". *Annual Review of Earth and Planetary Sciences*. **48**: 465–489. Bibcode:2020AREPS..48..465S. doi:10.1146/annurev-earth-081619-052855. S2CID 212832169. Archived from the original on May 7, 2020. Retrieved March 18,2022.

53. ^ Ingersoll, A. P.; Hammel, H. B.; Spilker, T. R.; Young, R. E. (June 1, 2005). "Outer Planets: The Ice Giants" (PDF). Lunar & Planetary Institute. Retrieved February 1, 2007.

54. ^ MacDougal, Douglas W. (2012). "A Binary System Close to Home: How the Moon and Earth Orbit Each Other". *Newton's Gravity*. Undergraduate Lecture Notes in Physics. Springer New York. pp. 193–211. doi:10.1007/978-1-4614-5444-1_10. ISBN 978-1-4614-5443-4. the barycentre is

743,000 km from the centre of the Sun. The Sun's radius is 696,000 km, so it is 47,000 km above the surface.

55. ^ a b c d e Burgess, Eric (1982). *By Jupiter: Odysseys to a Giant*. New York: Columbia University Press. ISBN 978-0-231-05176-7.

56. ^ Shu, Frank H. (1982). *The physical universe: an introduction to astronomy*. Series of books in astronomy (12th ed.). University Science Books. p. 426. ISBN 978-0-935702-05-7.

57. ^ Davis, Andrew M.; Turekian, Karl K. (2005). *Meteorites, comets, and planets*. Treatise on geochemistry. Vol. 1. Elsevier. p. 624. ISBN 978-0-08-044720-9.

58. ^ Schneider, Jean (2009). "The Extrasolar Planets Encyclopedia: Interactive Catalogue". Paris Observatory.

59. ^ Seager, S.; Kuchner, M.; Hier-Majumder, C. A.; Militzer, B. (2007). "Mass-Radius Relationships for Solid Exoplanets". *The Astrophysical Journal*. **669** (2): 1279–1297. arXiv:0707.2895. Bibcode:2007ApJ...669.1279S. doi:10.1086/521346. S2CID 8369390.

60. ^ a b c d e f g h *How the Universe Works 3*. Vol. Jupiter: Destroyer or Savior?. Discovery Channel. 2014.

61. ^ Guillot, Tristan (1999). "Interiors of Giant Planets Inside and Outside the Solar System"(PDF). *Science*. **286** (5437): 72–77. Bibcode:1999Sci...286...72G. doi:10.1126/science.286.5437.72. PMID 10506563. Retrieved April 24, 2022.

62. ^ Burrows, Adam; Hubbard, W. B.; Lunine, J. I.; Liebert, James (July 2001). "The theory of brown dwarfs and

extrasolar giant planets". *Reviews of Modern Physics*. **73** (3): 719–765. arXiv:astro-ph/0103383. Bibcode:2001RvMP...73..719B. doi:10.1103/RevModPhys.73.719. S2CID 204927572. Hence the HBMM at solar metallicity and $Y_\alpha = 50.25$ is $0.07 - 0.074$ $M_\odot$, ... while the HBMM at zero metallicity is 0.092 $M_\odot$

63. ^ von Boetticher, Alexander; Triaud, Amaury H. M. J.; Queloz, Didier; Gill, Sam; Lendl, Monika; Delrez, Laetitia; Anderson, David R.; Collier Cameron, Andrew; Faedi, Francesca; Gillon, Michaël; Gómez Maqueo Chew, Yilen; Hebb, Leslie; Hellier, Coel; Jehin, Emmanuël; Maxted, Pierre F. L.; Martin, David V.; Pepe, Francesco; Pollacco, Don; Ségransan, Damien; Smalley, Barry; Udry, Stéphane; West, Richard (August 2017). "The EBLM project. III. A Saturn-size low-mass star at the hydrogen-burning limit". *Astronomy & Astrophysics*. **604**: 6. arXiv:1706.08781. Bibcode:2017A&A...604L...6V. doi:10.1051/0004-6361/201731107. S2CID 54610182. L6.

64. ^ a b c d e f g h Elkins-Tanton, Linda T. (2011). *Jupiter and Saturn* (revised ed.). New York: Chelsea House. ISBN 978-0-8160-7698-7.

65. ^ Irwin, Patrick (2003). *Giant Planets of Our Solar System: Atmospheres, Composition, and Structure*. Springer Science & Business Media. p. 62. ISBN 9783540006817.

66. ^ Irwin, Patrick G. J. (2009) [2003]. *Giant Planets of Our Solar System: Atmospheres, Composition, and Structure* (Second ed.). Springer. p. 4. ISBN 978-3-642-09888-8. the

radius of Jupiter is estimated to be currently shrinking by approximately 1 mm/yr.

67. ^ a b Guillot, Tristan; Stevenson, David J.; Hubbard, William B.; Saumon, Didier (2004). "Chapter 3: The Interior of Jupiter". In Bagenal, Fran; Dowling, Timothy E.; McKinnon, William B. (eds.). *Jupiter: The Planet, Satellites and Magnetosphere*. Cambridge University Press. ISBN 978-0-521-81808-7.

68. ^ Bodenheimer, P. (1974). "Calculations of the early evolution of Jupiter". *Icarus*. 23. **23** (3): 319–325. Bibcode:1974Icar...23..319B. doi:10.1016/0019-1035(74)90050-5.

69. ^ Smoluchowski, R. (1971). "Metallic interiors and magnetic fields of Jupiter and Saturn". *The Astrophysical Journal*. **166**: 435. Bibcode:1971ApJ...166..435S. doi:10.1086/150971.

70. ^ Wall, Mike (May 26, 2017). "More Jupiter Weirdness: Giant Planet May Have Huge, 'Fuzzy' Core". space.com. Retrieved April 20, 2021.

71. ^ Weitering, Hanneke (January 10, 2018). "'Totally Wrong' on Jupiter: What Scientists Gleaned from NASA's Juno Mission". space.com. Retrieved February 26, 2021.

72. ^ Liu, S. F.; Hori, Y.; Müller, S.; Zheng, X.; Helled, R.; Lin, D.; Isella, A. (2019). "The formation of Jupiter's diluted core by a giant impact". *Nature*. **572** (7769): 355–357. arXiv:2007.08338. Bibcode:2019Natur.572..355L. doi:10.1038/s41586-019-1470-2. PMID 31413376. S2CID 199576704.

73. ^ Guillot, T. (2019). "Signs that Jupiter was mixed by a giant impact". *Nature*. **572** (7769): 315–317. Bibcode:2019Natur.572..315G. doi:10.1038/d41586-019-02401-1. PMID 31413374.

74. ^ Wahl, S. M.; Hubbard, William B.; Militzer, B.; Guillot, Tristan; Miguel, Y.; Movshovitz, N.; Kaspi, Y.; Helled, R.; Reese, D.; Galanti, E.; Levin, S.; Connerney, J. E.; Bolton, S. J. (2017). "Comparing Jupiter interior structure models to Juno gravity measurements and the role of a dilute core". *Geophysical Research Letters*. **44** (10): 4649–4659. arXiv:1707.01997. Bibcode:2017GeoRL..44.4649W. doi:10.1002/2017GL073160.

75. ^ Trachenko, K.; Brazhkin, V. V.; Bolmatov, D. (March 2014). "Dynamic transition of supercritical hydrogen: Defining the boundary between interior and atmosphere in gas giants". *Physical Review E*. **89** (3): 032126. arXiv:1309.6500. Bibcode:2014PhRvE..89c2126T. doi:10.1103/PhysRevE.89.032126. PMID 24730809. S2CID 42559818. 032126.

76. ^ "A Freaky Fluid inside Jupiter?". NASA. Retrieved December 8, 2021.

77. ^ "Oceans of diamond possible on Uranus and Neptune". Astronomy Now. Retrieved December 8, 2021.

78. ^ "NASA System Exploration Jupiter". NASA. Retrieved December 8, 2021.

79. ^ Guillot, T. (1999). "A comparison of the interiors of Jupiter and Saturn". *Planetary and Space Science*. **47** (10–11): 1183–1200. arXiv:astro-ph/9907402.

Bibcode:1999P&SS...47.1183G. doi:10.1016/S0032-0633(99)00043-4. S2CID 19024073.

80. ^ a b Lang, Kenneth R. (2003). "Jupiter: a giant primitive planet". NASA. Archived from the original on May 14, 2011. Retrieved January 10,2007.

81. ^ Lodders, Katharina (2004). "Jupiter Formed with More Tar than Ice" (PDF). *The Astrophysical Journal*. **611** (1): 587–597. Bibcode:2004ApJ...611..587L. doi:10.1086/421970. S2CID 59361587. Archived from the original (PDF) on April 12, 2020.

82. ^ S. Brygoo et al. 'Evidence of hydrogen–helium immiscibility at Jupiter-interior conditions.' *Nature.* Vol. 593, May 27, 2021, p. 517. doi:10.1038/s41586-021-03516-0.

83. ^ Kramer, Miriam (October 9, 2013). "Diamond Rain May Fill Skies of Jupiter and Saturn". Space.com. Retrieved August 27, 2017.

84. ^ Kaplan, Sarah (August 25, 2017). "It rains solid diamonds on Uranus and Neptune". The Washington Post. Retrieved August 27, 2017.

85. ^ a b Guillot, Tristan; Stevenson, David J.; Hubbard, William B.; Saumon, Didier (2004). "The interior of Jupiter". In Bagenal, Fran; Dowling, Timothy E.; McKinnon, William B. (eds.). *Jupiter. The planet, satellites and magnetosphere.* Cambridge planetary science. Vol. 1. Cambridge, UK: Cambridge University Press. p. 45. Bibcode:2004jpsm.book...35G. ISBN 0-521-81808-7.

86. ^ Loeffler, Mark J.; Hudson, Reggie L. (March 2018). "Coloring Jupiter's clouds: Radiolysis of ammonium hydrosulfide (NH4SH)" (PDF). *Icarus*. **302**: 418–425. Bibcode:2018Icar..302..418L. doi:10.1016/j.icarus.2017.10.041. Retrieved April 25, 2022.

87. ^ Ingersoll, Andrew P.; Dowling, Timothy E.; Gierasch, Peter J.; Orton, Glenn S.; Read, Peter L.; Sánchez-Lavega, Agustin; Showman, Adam P.; Simon-Miller, Amy A.; Vasavada, Ashwin R. (2004). Bagenal, Fran; Dowling, Timothy E.; McKinnon, William B. (eds.). "Dynamics of Jupiter's Atmosphere" (PDF). *Jupiter. The Planet, Satellites and Magnetosphere*. Cambridge planetary science. Cambridge, UK: Cambridge University Press. **1**: 105–128. ISBN 0-521-81808-7. Retrieved March 8, 2022.

88. ^ Aglyamov, Yury S.; Lunine, Jonathan; Becker, Heidi N.; Guillot, Tristan; Gibbard, Seran G.; Atreya, Sushil; Bolton, Scott J.; Levin, Steven; Brown, Shannon T.; Wong, Michael H. (February 2021). "Lightning Generation in Moist Convective Clouds and Constraints on the Water Abundance in Jupiter". *Journal of Geophysical Research: Planets*. **126** (2). arXiv:2101.12361. Bibcode:2021JGRE..12606504A. doi:10.1029/2020JE006504. S2CID 231728590. e06504.

89. ^ Watanabe, Susan, ed. (February 25, 2006). "Surprising Jupiter: Busy Galileo spacecraft showed jovian system is full of surprises". NASA. Retrieved February 20, 2007.

90. ^ Kerr, Richard A. (2000). "Deep, Moist Heat Drives Jovian Weather". *Science*. **287** (5455): 946–947.

doi:10.1126/science.287.5455.946b. S2CID 129284864.
Retrieved April 26, 2022.

91. ^ Becker, Heidi N.; Alexander, James W.; Atreya, Sushil K.; Bolton, Scott J.; Brennan, Martin J.; Brown, Shannon T.; Guillaume, Alexandre; Guillot, Tristan; Ingersoll, Andrew P.; Levin, Steven M.; Lunine, Jonathan I.; Aglyamov, Yury S.; Steffes, Paul G. (2020). "Small lightning flashes from shallow electrical storms on Jupiter". *Nature.* **584** (7819): 55–58. Bibcode:2020Natur.584...55B. doi:10.1038/s41586-020-2532-1. ISSN 0028-0836. PMID 32760043. S2CID 220980694.

92. ^ Guillot, Tristan; Stevenson, David J.; Atreya, Sushil K.; Bolton, Scott J.; Becker, Heidi N.(2020). "Storms and the Depletion of Ammonia in Jupiter: I. Microphysics of "Mushballs"". *Journal of Geophysical Research: Planets.* **125**(8): e2020JE006403. arXiv:2012.14316. Bibcode:2020JGRE..12506403G. doi:10.1029/2020JE006404. S2CID 226194362.

93. ^ Giles, Rohini S.; Greathouse, Thomas K.; Bonfond, Bertrand; Gladstone, G. Randall; Kammer, Joshua A.; Hue, Vincent; Grodent, Denis C.; Gérard, Jean-Claude; Versteeg, Maarten H.; Wong, Michael H.; Bolton, Scott J.; Connerney, John E. P.; Levin, Steven M. (2020). "Possible Transient Luminous Events Observed in Jupiter's Upper Atmosphere". *Journal of Geophysical Research: Planets.* **125** (11): e06659. arXiv:2010.13740. Bibcode:2020JGRE..12506659G. doi:10.1029/2020JE006659. S2CID 225075904. e06659.

94. ^ Greicius, Tony, ed. (October 27, 2020). "Juno Data Indicates 'Sprites' or 'Elves' Frolic in Jupiter's Atmosphere". *NASA*. Retrieved December 30, 2020.

95. ^ Strycker, P. D.; Chanover, N.; Sussman, M.; Simon-Miller, A. (2006). *A Spectroscopic Search for Jupiter's Chromophores*. DPS meeting #38, #11.15. American Astronomical Society. Bibcode:2006DPS....38.1115S.

96. ^ a b c Gierasch, Peter J.; Nicholson, Philip D.(2004). "Jupiter". World Book @ NASA. Archived from the original on January 5, 2005. Retrieved August 10, 2006.

97. ^ Chang, Kenneth (December 13, 2017). "The Great Red Spot Descends Deep into Jupiter". The New York Times. Retrieved December 15,2017.

98. ^ Denning, William F. (1899). "Jupiter, early history of the great red spot on". Monthly Notices of the Royal Astronomical Society. **59**(10): 574–584. Bibcode:1899MNRAS..59..574D. doi:10.1093/mnras/59.10.574.

99. ^ Kyrala, A. (1982). "An explanation of the persistence of the Great Red Spot of Jupiter". *Moon and the Planets*. **26** (1): 105–107. Bibcode:1982M&P....26..105K. doi:10.1007/BF00941374. S2CID 121637752.

100. ^ Oldenburg, Henry, ed. (1665–1666). "Philosophical Transactions of the Royal Society". Project Gutenberg. Retrieved December 22, 2011.

101. ^ Wong, M.; de Pater, I. (May 22, 2008). "New Red Spot Appears on Jupiter". *HubbleSite*. NASA. Retrieved December 12, 2013.

102.   ^ Simon-Miller, A.; Chanover, N.; Orton, G. (July 17, 2008). "Three Red Spots Mix It Up on Jupiter". *HubbleSite*. NASA. Retrieved April 26, 2015.

103.   ^ Covington, Michael A. (2002). *Celestial Objects for Modern Telescopes*. Cambridge University Press. p. 53. ISBN 978-0-521-52419-3.

104.   ^ Cardall, C. Y.; Daunt, S. J. "The Great Red Spot". University of Tennessee. Retrieved February 2, 2007.

105.   ^ *Jupiter, the Giant of the Solar System*. *The Voyager Mission*. NASA. 1979. p. 5.

106.   ^ Sromovsky, L. A.; Baines, K. H.; Fry, P. M.; Carlson, R. W. (July 2017). "A possibly universal red chromophore for modeling colour variations on Jupiter". *Icarus*. **291**: 232–244. arXiv:1706.02779. Bibcode:2017Icar..291..232S. doi:10.1016/j.icarus.2016.12.014. S2CID 119036239.

107.   ^ a b White, Greg (November 25, 2015). "Is Jupiter's Great Red Spot nearing its twilight?". *Space.news*. Retrieved April 13, 2017.

108.   ^ Sommeria, Jöel; Meyers, Steven D.; Swinney, Harry L. (February 25, 1988). "Laboratory simulation of Jupiter's Great Red Spot". *Nature*. **331** (6158): 689–693. Bibcode:1988Natur.331..689S. doi:10.1038/331689a0. S2CID 39201626.

109.   ^ a b Simon, A. A.; Wong, M. H.; Rogers, J. H.; Orton, G. S.; de Pater, I.; Asay-Davis, X.; Carlson, R. W.; Marcus, P. S. (March 2015). *Dramatic Change in Jupiter's Great Red Spot*. 46th Lunar and Planetary Science

Conference. March 16–20, 2015. The Woodlands, Texas. Bibcode:2015LPI....46.1010S.

110.   ^ Doctor, Rina Marie (October 21, 2015). "Jupiter's Superstorm Is Shrinking: Is Changing Red Spot Evidence Of Climate Change?". *Tech Times*. Retrieved April 13, 2017.

111.   ^ Grush, Loren (October 28, 2021). "NASA's *Juno*spacecraft finds just how deep Jupiter's Great Red Spot goes". *The Verge*. Retrieved October 28, 2021.

112.   ^ Adriani, A.; Mura, A.; Orton, G.; Hansen, C.; Altieri, F.; Moriconi, M. L.; Rogers, J.; Eichstädt, G.; Momary, T.; Ingersoll, A. P.; Filacchione, G.; Sindoni, G.; Tabataba-Vakili, F.; Dinelli, B. M.; Fabiano, F.; Bolton, S. J.; Connerney, J. E. P.; Atreya, S. K.; Lunine, J. I.; Tosi, F.; Migliorini, A.; Grassi, D.; Piccioni, G.; Noschese, R.; Cicchetti, A.; Plainaki, C.; Olivieri, A.; O'Neill, M. E.; Turrini, D.; Stefani, S.; Sordini, R.; Amoroso, M. (March 2018). "Clusters of cyclones encircling Jupiter's poles". *Nature*. **555** (7695): 216–219. Bibcode:2018Natur.555..216A. doi:10.1038/nature25491. PMID 29516997. S2CID 4438233.

113.   ^ Starr, Michelle (December 13, 2017). "NASA Just Watched a Mass of Cyclones on Jupiter Evolve Into a Mesmerising Hexagon". *Science Alert*.

114.   ^ Steigerwald, Bill (October 14, 2006). "Jupiter's Little Red Spot Growing Stronger". NASA. Retrieved February 2, 2007.

115.   ^ Wong, Michael H.; de Pater, Imke; Asay-Davis, Xylar; Marcus, Philip S.; Go, Christopher Y. (September

2011). "Vertical structure of Jupiter's Oval BA before and after it reddened: What changed?" (PDF). *Icarus*. **215** (1): 211–225. Bibcode:2011Icar..215..211W. doi:10.1016/j.icarus.2011.06.032. Retrieved April 27, 2022.

116. ^ Stallard, Tom S.; Melin, Henrik; Miller, Steve; Moore, Luke; O'Donoghue, James; Connerney, John E. P.; Satoh, Takehiko; West, Robert A.; Thayer, Jeffrey P.; Hsu, Vicki W.; Johnson, Rosie E. (April 10, 2017). "The Great Cold Spot in Jupiter's upper atmosphere". *Geophysical Research Letters*. **44** (7): 3000–3008. Bibcode:2017GeoRL..44.3000S. doi:10.1002/2016GL071956. PMC 5439487. PMID 28603321.

117. ^ Connerney, J. E. P.; Kotsiaros, S.; Oliversen, R. J.; Espley, J. R.; Joergensen, J. L.; Joergensen, P. S.; Merayo, J. M. G.; Herceg, M.; Bloxham, J.; Moore, K. M.; Bolton, S. J.; Levin, S. M. (May 26, 2017). "A New Model of Jupiter's Magnetic Field From Juno's First Nine Orbits" (PDF). *Geophysical Research Letters*. **45** (6): 2590–2596. Bibcode:2018GeoRL..45.2590C. doi:10.1002/2018GL077312.

118. ^ Brainerd, Jim (November 22, 2004). "Jupiter's Magnetosphere". *The Astrophysics Spectator*. Archived from the original on January 25, 2021. Retrieved August 10, 2008.

119. ^ "Receivers for Radio JOVE". *NASA*. March 1, 2017. Archived from the original on January 26, 2021. Retrieved September 9, 2020.

120.   ^ Phillips, Tony; Horack, John M. (February 20, 2004). "Radio Storms on Jupiter". *NASA*. Archived from the original on February 13, 2007. Retrieved February 1, 2007.

121.   ^ Showalter, M. A.; Burns, J. A.; Cuzzi, J. N.; Pollack, J. B. (1987). "Jupiter's ring system: New results on structure and particle properties". *Icarus*. **69** (3): 458–498. Bibcode:1987Icar...69..458S. doi:10.1016/0019-1035(87)90018-2.

122.   ^ a b Burns, J. A.; Showalter, M. R.; Hamilton, D. P.; Nicholson, P. D.; de Pater, I.; Ockert-Bell, M. E.; Thomas, P. C. (1999). "The Formation of Jupiter's Faint Rings". *Science*. **284** (5417): 1146–1150. Bibcode:1999Sci...284.1146B. doi:10.1126/science.284.5417.1146. PMID 10325220. S2CID 21272762.

123.   ^ Fieseler, P. D.; Adams, O. W.; Vandermey, N.; Theilig, E. E.; Schimmels, K. A.; Lewis, G. D.; Ardalan, S. M.; Alexander, C. J. (2004). "The Galileo Star Scanner Observations at Amalthea". *Icarus*. **169** (2): 390–401. Bibcode:2004Icar..169..390F. doi:10.1016/j.icarus.2004.01.012.

124.   ^ Herbst, T. M.; Rix, H.-W. (1999). "Star Formation and Extrasolar Planet Studies with Near-Infrared Interferometry on the LBT". In Guenther, Eike; Stecklum, Bringfried; Klose, Sylvio (eds.). *Optical and Infrared Spectroscopy of Circumstellar Matter*. ASP Conference Series. Vol. 188. San Francisco, Calif.: Astronomical Society of the Pacific. pp. 341–350. Bibcode:1999ASPC..188..341H. ISBN 978-1-58381-014-9. – See section 3.4.

125.	^ MacDougal, Douglas W. (December 16, 2012). *Newton's Gravity: An Introductory Guide to the Mechanics of the Universe*. Springer New York. p. 199. ISBN 978-1-4614-5444-1.

126.	^ *Popular Astronomy*. Vol. 44. Carleton College. 1936. p. 542.

127.	^ Michtchenko, T. A.; Ferraz-Mello, S. (February 2001). "Modeling the 5:2 Mean-Motion Resonance in the Jupiter–Saturn Planetary System". *Icarus*. **149** (2): 77–115. Bibcode:2001Icar..149..357M. doi:10.1006/icar.2000.6539.

128.	^ "Interplanetary Seasons". Science@NASA. Archived from the original on October 16, 2007. Retrieved February 20, 2007.

129.	^ Ridpath, Ian (1998). *Norton's Star Atlas*(19th ed.). Prentice Hall. ISBN 978-0-582-35655-9.[page needed]

130.	^ Hide, R. (January 1981). "On the rotation of Jupiter". *Geophysical Journal*. **64**: 283–289. Bibcode:1981GeoJ...64..283H. doi:10.1111/j.1365-246X.1981.tb02668.x.

131.	^ Russell, C. T.; Yu, Z. J.; Kivelson, M. G. (2001). "The rotation period of Jupiter" (PDF). *Geophysical Research Letters*. **28** (10): 1911–1912. Bibcode:2001GeoRL..28.1911R. doi:10.1029/2001GL012917. S2CID 119706637. Retrieved April 28, 2022.

132.	^ Rogers, John H. (July 20, 1995). "Appendix 3". *The giant planet Jupiter*. Cambridge University Press. ISBN 978-0-521-41008-3.

133. ^ Price, Fred W. (October 26, 2000). *The Planet Observer's Handbook*. Cambridge University Press. p. 140. ISBN 9780521789813.

134. ^ Fimmel, Richard O.; Swindell, William; Burgess, Eric (1974). "8. Encounter with the Giant". *Pioneer Odyssey* (Revised ed.). NASA History Office. Retrieved February 17,2007.

135. ^ Chaple, Glenn F. (2009). Jones, Lauren V.; Slater, Timothy F. (eds.). *Outer Planets*. Greenwood Guides to the Universe. ABC-CLIO. p. 47. ISBN 9780313365713.

136. ^ North, Chris; Abel, Paul (October 31, 2013). *The Sky at Night: How to Read the Solar System*. Ebury Publishing. p. 183. ISBN 978-1-4481-4130-2.

137. ^ Sachs, A. (May 2, 1974). "Babylonian Observational Astronomy". Philosophical Transactions of the Royal Society of London. **276** (1257): 43–50 (see p. 44). Bibcode:1974RSPTA.276...43S. doi:10.1098/rsta.1974.0008. JSTOR 74273. S2CID 121539390.

138. ^ Dubs, Homer H. (1958). "The Beginnings of Chinese Astronomy". *Journal of the American Oriental Society*. **78** (4): 295–300. doi:10.2307/595793. JSTOR 595793.

139. ^ Chen, James L.; Chen, Adam (2015). *A Guide to Hubble Space Telescope Objects: Their Selection, Location, and Significance*. Springer International Publishing. p. 195. ISBN 9783319188720.

140.　　　‸ Seargent, David A. J. (September 24, 2010). "Facts, Fallacies, Unusual Observations, and Other Miscellaneous Gleanings". *Weird Astronomy: Tales of Unusual, Bizarre, and Other Hard to Explain Observations*. Astronomers' Universe. pp. 221–282. ISBN 978-1-4419-6424-3.

141.　　　‸ Xi, Z. Z. (1981). "The Discovery of Jupiter's Satellite Made by Gan-De 2000 Years Before Galileo". *Acta Astrophysica Sinica*. **1** (2): 87. Bibcode:1981AcApS...1...85X.

142.　　　‸ Dong, Paul (2002). *China's Major Mysteries: Paranormal Phenomena and the Unexplained in the People's Republic*. China Books. ISBN 978-0-8351-2676-2.

143.　　　‸ Ossendrijver, Mathieu (January 29, 2016). "Ancient Babylonian astronomers calculated Jupiter's position from the area under a time-velocity graph". *Science*. **351** (6272): 482–484. Bibcode:2016Sci...351..482O. doi:10.1126/science.aad8085. PMID 26823423. S2CID 206644971.

144.　　　‸ Pedersen, Olaf (1974). *A Survey of the Almagest*. Odense University Press. pp. 423, 428. ISBN 9788774920878.

145.　　　‸ Pasachoff, Jay M. (2015). "Simon Marius's Mundus Iovialis: 400th Anniversary in Galileo's Shadow". *Journal for the History of Astronomy*. **46** (2): 218–234. Bibcode:2015AAS...22521505P. doi:10.1177/0021828615585493. S2CID 120470649.

146.　　　‸ Westfall, Richard S. "Galilei, Galileo". *The Galileo Project*. Rice University. Retrieved January 10, 2007.

147. 	^ O'Connor, J. J.; Robertson, E. F. (April 2003). "Giovanni Domenico Cassini". University of St. Andrews. Retrieved February 14, 2007.

148. 	^ Atkinson, David H.; Pollack, James B.; Seiff, Alvin (September 1998). "The Galileo probe Doppler wind experiment: Measurement of the deep zonal winds on Jupiter". *Journal of Geophysical Research*. **103** (E10): 22911–22928. Bibcode:1998JGR...10322911A. doi:10.1029/98JE00060.

149. 	^ Murdin, Paul (2000). *Encyclopedia of Astronomy and Astrophysics*. Bristol: Institute of Physics Publishing. ISBN 978-0-12-226690-4.

150. 	^ Rogers, John H. (1995). *The giant planet Jupiter*. Cambridge University Press. pp. 188–189. ISBN 9780521410083.

151. 	^ Fimmel, Richard O.; Swindell, William; Burgess, Eric (August 1974). "Jupiter, Giant of the Solar System". *Pioneer Odyssey*(Revised ed.). NASA History Office. Retrieved August 10, 2006.

152. 	^ Brown, Kevin (2004). "Roemer's Hypothesis". MathPages. Retrieved January 12, 2007.

153. 	^ Bobis, Laurence; Lequeux, James (July 2008). "Cassini, Rømer, and the velocity of light". *Journal of Astronomical History and Heritage*. **11**(2): 97–105. Bibcode:2008JAHH...11...97B.

154. 	^ Tenn, Joe (March 10, 2006). "Edward Emerson Barnard". Sonoma State University. Archived from the original on September 17, 2011. Retrieved January 10, 2007.

155.   ^ "Amalthea Fact Sheet". NASA/JPL. October 1, 2001. Archived from the original on November 24, 2001. Retrieved February 21,2007.

156.   ^ Dunham Jr., Theodore (1933). "Note on the Spectra of Jupiter and Saturn". *Publications of the Astronomical Society of the Pacific*. **45**(263): 42–44. Bibcode:1933PASP...45...42D. doi:10.1086/124297.

157.   ^ Youssef, A.; Marcus, P. S. (2003). "The dynamics of jovian white ovals from formation to merger". *Icarus*. **162** (1): 74–93. Bibcode:2003Icar..162...74Y. doi:10.1016/S0019-1035(02)00060-X.

158.   ^ Chang, Kenneth (July 15, 2022). "NASA Shows Webb's View of Something Closer to Home: Jupiter - The powerful telescope will help scientists make discoveries both within our solar system and well beyond it". The New York Times. Retrieved July 16, 2022.

159.   ^ Weintraub, Rachel A. (September 26, 2005). "How One Night in a Field Changed Astronomy". NASA. Retrieved February 18,2007.

160.   ^ Garcia, Leonard N. "The Jovian Decametric Radio Emission". NASA. Retrieved February 18, 2007.

161.   ^ Klein, M. J.; Gulkis, S.; Bolton, S. J. (1996). "Jupiter's Synchrotron Radiation: Observed Variations Before, During and After the Impacts of Comet SL9". *Conference at University of Graz*. NASA: 217. Bibcode:1997pre4.conf..217K. Retrieved February 18, 2007.

162.   ^ "The Pioneer Missions". NASA. March 26, 2007. Retrieved February 26, 2021.

163.   ^ "NASA Glenn Pioneer Launch History". NASA – Glenn Research Center. March 7, 2003. Retrieved December 22, 2011.

164.   ^ Fortescue, Peter W.; Stark, John; Swinerd, Graham (2003). *Spacecraft systems engineering*(3rd ed.). John Wiley and Sons. p. 150. ISBN 978-0-470-85102-9.

165.   ^ Hirata, Chris. "Delta-V in the Solar System". California Institute of Technology. Archived from the original on July 15, 2006. Retrieved November 28, 2006.

166.   ^ Wong, Al (May 28, 1998). "Galileo FAQ: Navigation". NASA. Archived from the original on January 5, 1997. Retrieved November 28, 2006.

167.   ^ a b c Chan, K.; Paredes, E. S.; Ryne, M. S. (2004). "Ulysses Attitude and Orbit Operations: 13+ Years of International Cooperation". *Space OPS 2004 Conference*. American Institute of Aeronautics and Astronautics. doi:10.2514/6.2004-650-447.

168.   ^ Lasher, Lawrence (August 1, 2006). "Pioneer Project Home Page". NASA Space Projects Division. Archived from the original on January 1, 2006. Retrieved November 28, 2006.

169.   ^ "Jupiter". NASA/JPL. January 14, 2003. Retrieved November 28, 2006.

170.   ^ Hansen, C. J.; Bolton, S. J.; Matson, D. L.; Spilker, L. J.; Lebreton, J.-P. (2004). "The Cassini–Huygens flyby of Jupiter". *Icarus*. **172**(1): 1–8. Bibcode:2004Icar..172....1H. doi:10.1016/j.icarus.2004.06.018.

171.	^ "Pluto-Bound New Horizons Sees Changes in Jupiter System". NASA. October 9, 2007. Retrieved February 26, 2021.

172.	^ "Pluto-Bound New Horizons Provides New Look at Jupiter System". NASA. May 1, 2007. Retrieved July 27, 2007.

173.	^ a b McConnell, Shannon (April 14, 2003). "Galileo: Journey to Jupiter". NASA/JPL. Archived from the original on November 3, 2004. Retrieved November 28, 2006.

174.	^ Magalhães, Julio (December 10, 1996). "Galileo Probe Mission Events". NASA Space Projects Division. Archived from the originalon January 2, 2007. Retrieved February 2, 2007.

175.	^ Goodeill, Anthony (March 31, 2008). "New Frontiers – Missions – Juno". NASA. Archived from the original on February 3, 2007. Retrieved January 2, 2007.

176.	^ "Juno, NASA's Jupiter probe". The Planetary Society. Retrieved April 27, 2022.

177.	^ Jet Propulsion Laboratory (June 17, 2016). "NASA's Juno spacecraft to risk Jupiter's fireworks for science". *phys.org*. Retrieved April 10, 2022.

178.	^ Firth, Niall (September 5, 2016). "NASA's Juno probe snaps first images of Jupiter's north pole". *New Scientist*. Retrieved September 5,2016.

179.	^ Clark, Stephen (February 21, 2017). "NASA's Juno spacecraft to remain in current orbit around Jupiter". Spaceflight Now. Retrieved April 26, 2017.

180.   ^ Agle, D. C.; Wendel, JoAnna; Schmid, Deb (June 6, 2018). "NASA Re-plans Juno's Jupiter Mission". NASA/JPL. Retrieved January 5,2019.

181.   ^ Talbert, Tricia (January 8, 2021). "NASA Extends Exploration for Two Planetary Science Missions". *NASA*. Retrieved January 11, 2021.

182.   ^ Dickinson, David (February 21, 2017). "Juno Will Stay in Current Orbit Around Jupiter". *Sky & Telescope*. Retrieved January 7, 2018.

183.   ^ Bartels, Meghan (July 5, 2016). "To protect potential alien life, NASA will destroy its $1 billion Jupiter spacecraft on purpose". *Business Insider*. Retrieved January 7, 2018.

184.   ^ Berger, Brian (February 7, 2005). "White House scales back space plans". MSNBC. Retrieved January 2, 2007.

185.   ^ "Laplace: A mission to Europa & Jupiter system". European Space Agency. Retrieved January 23, 2009.

186.   ^ Favata, Fabio (April 19, 2011). "New approach for L-class mission candidates". European Space Agency.

187.   ^ "ESA Science & Technology – JUICE". ESA. November 8, 2021. Retrieved November 10,2021.

188.   ^ Foust, Jeff (July 10, 2020). "Cost growth prompts changes to Europa Clipper instruments". *Space News*. Retrieved July 10,2020.

189.   ^ Jones, Andrew (January 12, 2021). "Jupiter Mission by China Could Include Callisto Landing". The Planetary Society. Retrieved April 27, 2020.

190.   ^ Jones, Andrew (April 16, 2021). "China to launch a pair of spacecraft towards the edge of the solar system". *Space News*. Retrieved April 27, 2020.

191.   ^ Billings, Lee (November 12, 2019). "Proposed Interstellar Mission Reaches for the Stars, One Generation at a Time". *Scientific American*. Retrieved April 27, 2020.

192.   ^ Sheppard, Scott S. "The Giant Planet Satellite and Moon Page". *Department of Terrestrial Magnetism*. Carnegie Institution for Science. Archived from the original on June 7, 2009. Retrieved December 19, 2014.

193.   ^ Zimmermann, Kim Ann (October 1, 2018). "Jupiter's Moons: Facts About the Largest Jovian Moons". *Space.com*. Retrieved December 31, 2020.

194.   ^ Carter, Jamie (2015). *A Stargazing Program for Beginners*. Springer International Publishing. p. 104. ISBN 978-3-319-22072-7.

195.   ^ Musotto, S.; Varadi, F.; Moore, W. B.; Schubert, G. (2002). "Numerical simulations of the orbits of the Galilean satellites". *Icarus*. **159** (2): 500–504. Bibcode:2002Icar..159..500M. doi:10.1006/icar.2002.6939.

196.   ^ a b Lang, Kenneth R. (March 3, 2011). *The Cambridge Guide to the Solar System*. Cambridge University Press. p. 304. ISBN 978-1-139-49417-5.

197.   ^ McFadden, Lucy-Ann; Weissmann, Paul; Johnson, Torrence (2006). *Encyclopedia of the Solar System*. Elsevier Science. p. 446. ISBN 978-0-08-047498-4.

198.   ^ Kessler, Donald J. (October 1981). "Derivation of the collision probability between orbiting objects: the

lifetimes of jupiter's outer moons". *Icarus*. **48** (1): 39–48. Bibcode:1981Icar...48...39K. doi:10.1016/0019-1035(81)90151-2. Retrieved December 30,2020.

199.   ^ Hamilton, Thomas W. M. (2013). *Moons of the Solar System*. SPBRA. p. 14. ISBN 978-1-62516-175-8.

200.   ^ Jewitt, D. C.; Sheppard, S.; Porco, C. (2004). Bagenal, F.; Dowling, T.; McKinnon, W. (eds.). *Jupiter: The Planet, Satellites and Magnetosphere* (PDF). Cambridge University Press. ISBN 978-0-521-81808-7. Archived from the original (PDF) on March 26, 2009.

201.   ^ a b c Nesvorný, D.; Alvarellos, J. L. A.; Dones, L.; Levison, H. F. (2003). "Orbital and Collisional Evolution of the Irregular Satellites" (PDF). *The Astronomical Journal*. **126** (1): 398–429. Bibcode:2003AJ....126..398N. doi:10.1086/375461.

202.   ^ Showman, A. P.; Malhotra, R. (1999). "The Galilean Satellites". *Science*. **286** (5437): 77–84. Bibcode:1999Sci...296...77S. doi:10.1126/science.286.5437.77. PMID 10506564. S2CID 9492520.

203.   ^ Sheppard, Scott S.; Jewitt, David C. (May 2003). "An abundant population of small irregular satellites around Jupiter" (PDF). *Nature*. **423** (6937): 261–263. Bibcode:2003Natur.423..261S. doi:10.1038/nature01584. PMID 12748634. S2CID 4424447. Archived from the original(PDF) on August 13, 2006.

204.   ^ Nesvorný, David; Beaugé, Cristian; Dones, Luke; Levison, Harold F. (July 2003). "Collisional Origin of

Families of Irregular Satellites" (PDF). *The Astronomical Journal*. **127** (3): 1768–1783. Bibcode:2004AJ....127.1768N. doi:10.1086/382099.

205. ^ Ferraz-Mello, S. (1994). Milani, Andrea; Di Martino, Michel; Cellino, A. (eds.). *Kirkwood Gaps and Resonant Groups*. Asteroids, Comets, Meteors 1993: Proceedings of the 160th Symposium of the International Astronomical Union, held in Belgirate, Italy, June 14–18, 1993, International Astronomical Union. Symposium no. 160. Dordrecht: Kluwer Academic Publishers. p. 175. Bibcode:1994IAUS..160..175F.

206. ^ Kerr, Richard A. (2004). "Did Jupiter and Saturn Team Up to Pummel the Inner Solar System?". *Science*. **306** (5702): 1676. doi:10.1126/science.306.5702.1676a. PMID 15576586. S2CID 129180312.

207. ^ "List Of Jupiter Trojans". *IAU Minor Planet Center*. Retrieved October 24, 2010.

208. ^ Cruikshank, D. P.; Dalle Ore, C. M.; Geballe, T. R.; Roush, T. L.; Owen, T. C.; Cash, Michele; de Bergh, C.; Hartmann, W. K. (October 2000). "Trojan Asteroid 624 Hektor: Constraints on Surface Composition". *Bulletin of the American Astronomical Society*. **32**: 1027. Bibcode:2000DPS....32.1901C.

209. ^ Quinn, T.; Tremaine, S.; Duncan, M. (1990). "Planetary perturbations and the origins of short-period comets". *Astrophysical Journal, Part 1*. **355**: 667–679. Bibcode:1990ApJ...355..667Q. doi:10.1086/168800.

210.   ^ "Caught in the act: Fireballs light up Jupiter". *ScienceDaily*. September 10, 2010. Retrieved April 26, 2022.

211.   ^ Nakamura, T.; Kurahashi, H. (1998). "Collisional Probability of Periodic Comets with the Terrestrial Planets: An Invalid Case of Analytic Formulation". *Astronomical Journal*. **115** (2): 848–854. Bibcode:1998AJ....115..848N. doi:10.1086/300206.

212.   ^ Horner, J.; Jones, B. W. (2008). "Jupiter – friend or foe? I: the asteroids". *International Journal of Astrobiology*. 7 (3–4): 251–261. arXiv:0806.2795. Bibcode:2008IJAsB...7..251H. doi:10.1017/S1473550408004187. S2CID 8870726.

213.   ^ Overbye, Dennis (July 25, 2009). "Jupiter: Our Comic Protector?". *The New York Times*. Retrieved July 27, 2009.

214.   ^ "In Depth | P/Shoemaker-Levy 9". *NASA Solar System Exploration*. Retrieved December 3, 2021.

215.   ^ Howell, Elizabeth (January 24, 2018). "Shoemaker-Levy 9: Comet's Impact Left Its Mark on Jupiter". *Space.com*. Retrieved December 3, 2021.

216.   ^ information@eso.org. "The Big Comet Crash of 1994 – Intensive Observational Campaign at ESO". *www.eso.org*. Retrieved December 3,2021.

217.   ^ "Top 20 Comet Shoemaker-Levy Images". *www2.jpl.nasa.gov*. Retrieved December 3,2021.

218.   ^ information@eso.org. "Comet P/Shoemaker-Levy 9 "Gang Of Four"". *www.spacetelescope.org*. Retrieved December 3, 2021.

219.   ^ Savage, Donald; Elliott, Jim; Villard, Ray (December 30, 2004). "Hubble Observations Shed New Light on Jupiter Collision". *nssdc.gsfc.nasa.gov*. Retrieved December 3,2021.

220.   ^ "NASA TV Coverage on Comet Shoemaker-Levy". *www2.jpl.nasa.gov*. Retrieved December 3, 2021.

221.   ^ Tabe, Isshi; Watanabe, Jun-ichi; Jimbo, Michiwo (February 1997). "Discovery of a Possible Impact SPOT on Jupiter Recorded in 1690". *Publications of the Astronomical Society of Japan*. **49**: L1–L5. Bibcode:1997PASJ...49L...1T. doi:10.1093/pasj/49.1.11.

222.   ^ "Stargazers prepare for daylight view of Jupiter". ABC News. June 16, 2005. Archived from the original on May 12, 2011. Retrieved February 28, 2008.

223.   ^ a b Rogers, J. H. (1998). "Origins of the ancient constellations: I. The Mesopotamian traditions". *Journal of the British Astronomical Association*. **108**: 9–28. Bibcode:1998JBAA..108....9R.

224.   ^ Waerden, B. L. (1974). "Old-Babylonian Astronomy" (PDF). *Science Awakening II*. Dordrecht: Springer: 46–59. doi:10.1007/978-94-017-2952-9_3. ISBN 978-90-481-8247-3. Retrieved March 21, 2022.

225.   ^ "Greek Names of the Planets". April 25, 2010. Retrieved July 14, 2012. In Greek the name of the planet Jupiter is Dias, the Greek name of god Zeus. See also the Greek article about the planet.

226.   ^ Cicero, Marcus Tullius (1888). *Cicero's Tusculan Disputations; also, Treatises on The Nature of the Gods, and*

on The Commonwealth. Translated by Yonge, Charles Duke. New York, NY: Harper & Brothers. p. 274 – via Internet Archive.

227.　　^ Cicero, Marcus Tullus (1967) [1933]. Warmington, E. H. (ed.). *De Natura Deorum*[*On The Nature of the Gods*]. Cicero. Vol. 19. Translated by Rackham, H. Cambridge, MA: Cambridge University Press. p. 175 – via Internet Archive.

228.　　^ Zolotnikova, O. (2019). "Mythologies in contact: Syro-Phoenician traits in Homeric Zeus". *The Scientific Heritage*. **41** (5): 16–24. Retrieved April 26, 2022.

229.　　^ Tarnas, R. (2009). "The planets". *Archai: The Journal of Archetypal Cosmology*. **1** (1): 36–49. CiteSeerX 10.1.1.456.5030. Retrieved March 21, 2022.

230.　　^ Harper, Douglas (November 2001). "Jupiter". *Online Etymology Dictionary*. Retrieved February 23, 2007.

231.　　^ "Guru". Indian Divinity.com. Retrieved February 14, 2007.

232.　　^ Sanathana, Y. S.; Manjil, Hazarika1 (November 27, 2020). "Astrolatry in the Brahmaputra Valley: Reflecting upon the Navagraha Sculptural Depiction" (PDF). *Heritage: Journal of Multidisciplinary Studies in Archaeology*. **8** (2): 157–174. Retrieved July 4, 2022.[dead link]

233.　　^ "Türk Astrolojisi-2" (in Turkish). NTV. Archived from the original on January 4, 2013. Retrieved April 23, 2010.

234.　　^ De Groot, Jan Jakob Maria (1912). *Religion in China: universism. a key to the study of Taoism and Confucianism*. *American lectures on the history of religions.*

Vol. 10. G.P. Putnam's Sons. p. 300. Retrieved January 8, 2010.

235.   ^ Crump, Thomas (1992). *The Japanese numbers game: the use and understanding of numbers in modern Japan*. Nissan Institute/Routledge Japanese studies series. Routledge. pp. 39–40. ISBN 978-0-415-05609-0.

236.   ^ Hulbert, Homer Bezaleel (1909). *The passing of Korea*. Doubleday, Page & Company. p. 426. Retrieved January 8, 2010.

237.   ^ Dubs, Homer H. (1958). "The Beginnings of Chinese Astronomy". *Journal of the American Oriental Society*. **78** (4): 295–300. doi:10.2307/595793. JSTOR 595793.

238.   ^ "By Jove! Jupiter Shows Its Stripes and Colors". *NOIRLab*. National Science Foundation. May 11, 2021. Retrieved June 17,2021.

239.   ^ "Hubble Finds Evidence of Persistent Water Vapour Atmosphere on Europa". *ESA Hubble*. European Space Agency. October 14, 202Y1. Retrieved October 26, 2021.

2

Fact About Jupiter

Ten Interesting Facts About Jupiter

Jupiter was appropriately named after the king of the gods. It's massive, has a powerful magnetic field, and more moons that any planet in the Solar System. Though it has been known to astronomers since ancient times, the invention of the telescope and the advent of modern astronomy has taught us so much about this gas giant.

In short, there are countless interesting facts about this gas giant that many people just don't know about. And we here at Universe Today have taken the liberty of compiling a list of ten particularly interesting ones that we think will fascinate and surprise you. Think you know everything about Jupiter? Think again!

1. Jupiter Is Massive:

It's no secret that Jupiter is the largest planet in the Solar System. But this description really doesn't do it justice. For one, the mass of Jupiter is 318 times as massive as the Earth. In fact, Jupiter is 2.5

times more massive than all of the other planets in the Solar System combined. But here's the really interesting thing…

Jupiter is the largest planet in our Solar System, with 2.5 times the mass of all the other planets combined.

If Jupiter got any more massive, it would actually get smaller. Additional mass would actually make the planet more dense, which would cause it to start pulling it in on itself. Astronomers estimate that Jupiter could end up with 4 times its current mass, and still remain about the same size.

2. Jupiter Cannot Become A Star:

Astronomers call Jupiter a failed star, but that's not really an appropriate description. While it is true that, like a star, Jupiter is rich in hydrogen and helium, Jupiter does not have nearly enough mass to trigger a fusion reaction in its core. This is how stars generate energy, by fusing hydrogen atoms together under extreme heat and pressure to create helium, releasing light and heat in the process.

This is made possible by their enormous gravity. For Jupiter to ignite a nuclear fusion process and become a star, it would need more than 70 times its current mass. If you could crash dozens of Jupiter together, you might have a chance to make a new star. But in the meantime, Jupiter shall remain a large gas giant with no hopes of becoming a star. Sorry, Jupiter!

3. Jupiter Is The Fastest Spinning Planet In The Solar System:

For all its size and mass, Jupiter sure moves quickly. In fact, with an rotational velocity of 12.6 km/s (~7.45 m/s) or 45,300 km/h(28,148 mph), the planet only takes about 10 hours to complete a full rotation on its axis. And because it's spinning so rapidly, the planet has flattened out at the poles a little and is bulging at its equator.

In fact, points on Jupiter's equator are more than 4,600 km further from the center than the poles. Or to put it another way, the planet's polar radius measures to 66,854 ± 10 km (or 10.517 that of Earth's), while its diameter at the equator is 71,492 ± 4 km (or 11.209 that of Earth's). This rapid rotation also helps generate Jupiter's powerful magnetic fields, and contribute to the dangerous radiation surrounding it.

4. The Clouds On Jupiter Are Only 50 km Thick:

That's right, all those beautiful whirling clouds and storms you see on Jupiter are only about 50 km thick. They're made of ammonia crystals broken up into two different cloud decks. The darker material is thought to be compounds brought up from deeper inside Jupiter, and then change color when they reacted with sunlight. But below those clouds, it's just hydrogen and helium, all the way down.

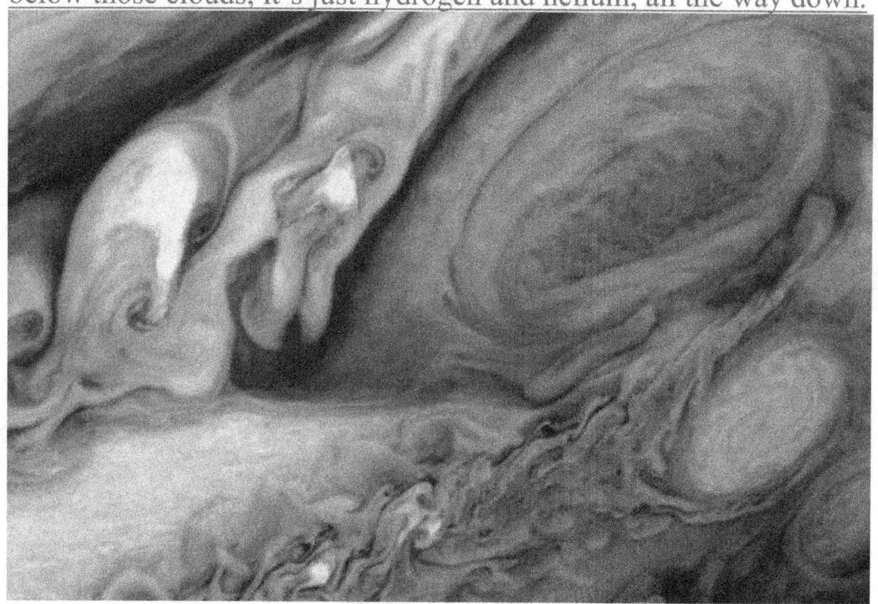

Image of Jupiter's Giant Red Spot, taken on March 5th, 1979.
Credit: NASA GSFC/NASA/JPL

5. The Great Red Spot Has Been Around For A Long Time:

The Great Red Spot on Jupiter is one of its most familiar features. This persistent anticyclonic storm, which is located south of its equator, measures between 24,000 km in diameter and 12–14,000 km in height. As such, it is large enough to contain two or three planets the size of Earth's diameter. And the spot has been around for at least 350 years, since it was spotted as far back as the 17th century.

The Great Red Spot was first identified in 1665 by Italian astronomer Giovanni Cassini. By the 20th century, astronomers began to theorize that it was a storm, one which was created by Jupiter's turbulent and fast-moving atmosphere. These theories were confirmed by the *Voyager 1* mission, which observed the Giant Red Spot up close in March of 1979 during its flyby of the planet.

However, it appears to have been shrinking since that time. Based on Cassini's observations, the size was estimated to be 40,000 km in the 17th century, which was almost twice as large as it is now. Astronomers do not know if or when it will ever disappear entirely, but they are relatively sure that another one will emerge somewhere else on the planet.

6. Jupiter Has Rings:

When people think of ring systems, Saturn naturally comes to mind. But in truth, both Uranus and Jupiter have ring systems of their own. Jupiter's were the third set to be discovered (after the other two), due to the fact that they are particularly faint. Jupiter's rings consist of three main segments – an inner torus of particles known as the halo, a relatively bright main ring, and an outer gossamer ring.

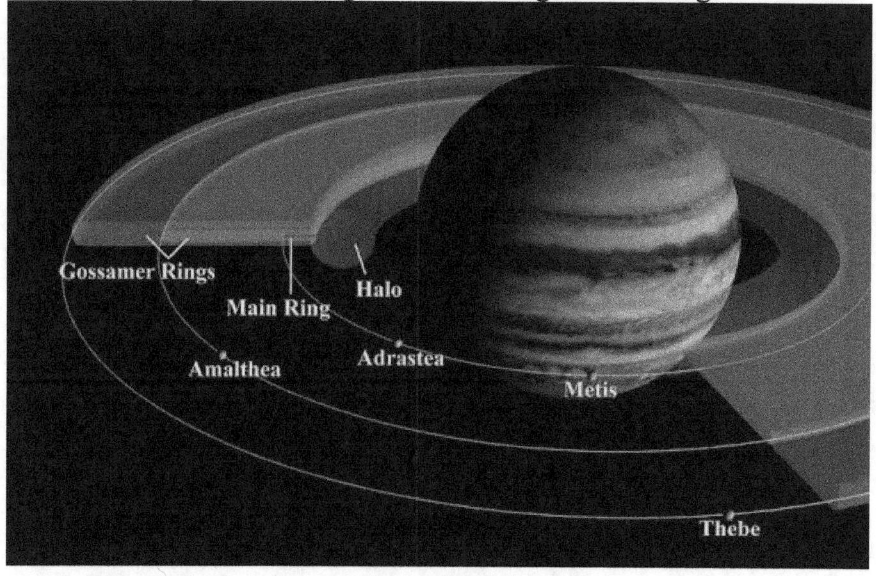

A schema of Jupiter's ring system showing the four main components. For simplicity, Metis and Adrastea are depicted as sharing their orbit.

These rings are widely believed to have come from material ejected by its moons when they're struck by meteorite impacts. In particular, the main ring is thought to be composed of material from the moons of Adrastea and Metis, while the moons of Thebe and Amalthea are believed to produce the two distinct components of the dusty gossamer ring.

This material fell into orbit around Jupiter (instead of falling back to their respective moons) because if Jupiter's strong gravitational influence. The ring is also depleted and replenished regularly as some material veers towards Jupiter while new material is added by additional impacts.

7. Jupiter's Magnetic Field Is 14 Times Stronger Than Earth's:

Compasses would really work on Jupiter. That's because it has the strongest magnetic field in the Solar System. Astronomers think the magnetic field is generated by the eddy currents – i.e. swirling movements of conducting materials – within the liquid metallic hydrogen core. This magnetic field traps particles of sulfur dioxide from Io's volcanic eruptions, which producing sulfur and oxygen ions. Together with hydrogen ions originating from the atmosphere of Jupiter, these form a plasma sheet in Jupiter's equatorial plane.

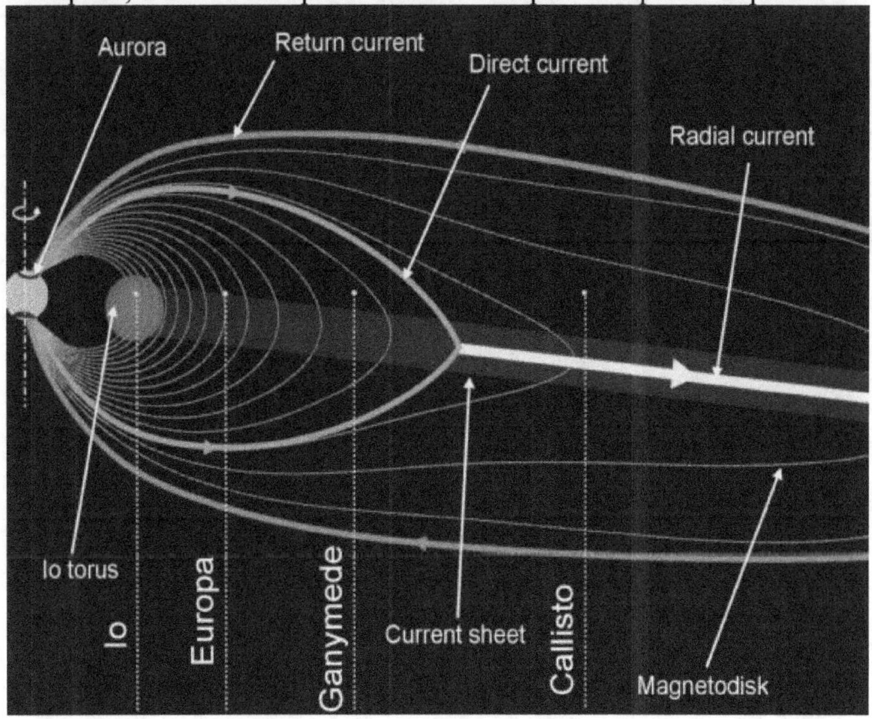

The magnetic field of Jupiter and co-rotation enforcing currents.

Credit.

Farther out, the interaction of the magnetosphere with the solar wind generates a bow shock, a dangerous belt of radiation that can cause damage tos spacecraft. Jupiter's four largest moons all orbit within the magnetosphere, which protects them from the solar wind, but also make the likelihood of establishing outposts on their surface problematic. The magnetosphere of Jupiter is also responsible for intense episodes of radio emission from the planet's polar regions.

8. Jupiter Has 67 Moons:

As of the penning of this article, Jupiter has a 67 confirmed and named satellites. However, it is estimated that the planet has over 200 natural satellites orbiting it. Almost all of them are less than 10 kilometers in diameter, and were only discovered after 1975, when the first spacecraft (*Pioneer 10*) arrived at Jupiter.

However, it also has four major moons, which are collectively known as the Galilean Moons(after their discovered Galileo Galilei). These are, in order of distance from Jupiter, Io, Europa, Ganymede, and Callisto. These moons are some of the largest in the Solar System, with Ganymede being the largest, measuring 5262 km in diameter.

Illustration of Jupiter and the Galilean satellites. Credit: NASA

9. Jupiter Has Been Visited 7 Times By Spacecraft:

Jupiter was first visited by NASA's *Pioneer 10*spacecraft in December 1973, and then *Pioneer 11* in December 1974. Then came the *Voyager 1 and2* flybys, both of which happened in 1979. This was followed by a long break until *Ulysses* arrived in February 1992, followed by the *Galileo* space probe in 1995. Then *Cassini* made a flyby in 2000, on its way to Saturn. And finally, NASA's _New Horizons_ spacecraft made its flyby in 2007. This was the last mission to fly past Jupiter, but it surely won't be the last.

10. You Can See Jupiter With Your Own Eyes:

Jupiter is the third brightest object in the Solar System, after Venus and the Moon. Chances are, you saw Jupiter in the sky, and had no idea that's what you were seeing. And here at Universe Today, we are in the habit of letting readers know when the best opportunities for spotting Jupiter in the night sky are.

Chances are, if you see a really bright star high in the sky, then you're looking at Jupiter. Get your hands on a pair of binoculars, and if you know someone with a telescope, that's even better. Using even modest magnification, you might even spot small specks of light orbiting it, which are its Galilean Moons. Just think, you'll be seeing precisely what Galileo did when he gazed at the planet in 1610.

www.ingramcontent.com/pod-product-compliance
Lightning Source LLC
Chambersburg PA
CBHW072103150726
47999CB00005B/1865